FRAGILE BEGINNINGS

Fragile Beginnings

Discoveries and Triumphs in the Newborn ICU

ADAM WOLFBERG, MD

A Harvard Health Publications Book
Beacon Press, Boston

Beacon Press
25 Beacon Street
Boston, Massachusetts 02108-2892
www.beacon.org

Beacon Press books
are published under the auspices of
the Unitarian Universalist Association of Congregations.

Printed in the United States of America

16 15 14 13 8 7 6 5 4 3 2 1

This book is printed on acid-free paper that meets the uncoated paper
ANSI/NISO specifications for permanence as revised in 1992.

Some names and identifying characteristics of people mentioned
in this work have been changed to protect their identities.

Text design by Wilsted & Taylor Publishing Services

Library of Congress Cataloging-in-Publication Data

Wolfberg, Adam.
Fragile beginnings : discoveries and triumphs in the newborn ICU /
Adam Wolfberg.
p. cm. — (A Harvard health publications book)
Includes bibliographical references and index.
ISBN 978-0-8070-1166-9 (paperback: alk. paper)
1. Neonatal intensive care—Popular works. 2. Medical ethics—
Popular works. I. Title.
RJ253.5.W65 2012
618.92'01—dc23 2011027177

To Kelly, whose fierce love of her family inspires all of us

Contents

CHAPTER I

Fragile Beginnings

A vaulted ceiling rises five stories over the lobby of the Brigham and Women's Hospital, its skylights illuminating the entrance to one of Harvard Medical School's flagship institutions. Approximately twenty-five times every day, a small but joyous parade descends from one of the postpartum floors and crosses the two hundred feet of polished tile that stretches from the maternity wing to the hospital entrance.

First comes an orderly pushing a wheelchair bearing a new mother who holds her swaddled newborn in her arms. Next comes another aide driving a cart piled high with flowers, cards, assorted baby gifts, and the usual infant paraphernalia: diapers, formula, a blanket, pacifiers. Last comes the beaming father, suitcase in hand, carrying whatever doesn't fit on the cart.

The procession pauses just before the revolving doors, and the father is dispatched to the parking garage. Soon the car pulls up, and the group moves out to the driveway; the new mother is lovingly strapped into the passenger seat and the baby carefully buckled into a car seat behind her. Quickly, luggage, gifts, and flowers are loaded, and the family drives away.

This is the way childbearing is supposed to happen.

Twice before, Kelly had ridden that wheelchair holding a healthy newborn. Understandably, she thought that having a child was a relatively uncomplicated affair: a couple tries to get pregnant and eventually succeeds. Nine months later, a healthy baby is born, and after a few days' recuperation, they go home together.

But after Larissa was born, Kelly rode the wheelchair across the lobby alone.

JANUARY 10

Kelly steered her Ford Explorer through the slow curve as Storrow Drive swung under the Boston University Bridge and entered the stretch of highway that she liked best in the morning. Next to her, the Charles River widened, and up ahead, the sun bounced off the skyscrapers and shimmered across the ice. Behind her, three-year-old Hannah, her younger child, was cheerfully holding up her end of a conversation.

"We have turtles, Mommy. We have turtles in a box, Mommy. It's my turn to feed the turtles, Mommy."

"What do you feed the turtles, Hanni?"

"Food, Mommy."

"What do turtles eat, Hanni?"

"They eat turtle food, Mommy."

"Of course they do, Hanni. I'll bet you are very good at feeding the turtles."

Before long, they arrived at Hannah's nursery school. Parking places were in short supply, so Kelly parked in the shopping plaza across the street. She had a long mental to-do list, and as she got Hannah out of the car, she added one more item—pick up some milk at the Stop & Shop before she left so she could get her ticket validated and exit for free.

"Hold my hand, love." They carefully crossed two lanes of Cambridge Street, squeezing into the space on the median strip between two snowbanks. There was a break in the traffic and they dashed across to the other side.

Kelly had come to enjoy the daily trip into Boston. The buildings, the traffic, and the streets lined with restaurants, dry cleaners, and coffee shops gave her a dose of urban living that she sometimes missed amid the setback suburban homes in her neighborhood with clipped lawns now covered by un-mussed snow.

Wrapped in layers against the January weather, they made their way to Hannah's classroom. At first, Hannah clung to Kelly's leg and pushed her face into Kelly's coat.

"Hello, Priscilla." Kelly greeted one of Hannah's favorite teachers warmly. "You have another boring day planned for the children? A whole lot of sitting around?"

"That's what we'll do today," Priscilla said, smiling. "A whole lot of sitting and staring out the window. We will do our best to have no fun at all."

Hannah's face emerged. She looked up at Priscilla, then at Kelly, then back at Priscilla, and started to smile. She loosened her grip.

"Want to read a book, Mommy?"

"You know it." They chose *Guess How Much I Love You* off the shelves and went to the reading area with its low couches and pillows. Awkwardly, Kelly lowered herself onto the couch. "Renee, Ella, Alex, do you want to read with us?" she asked some of Hannah's friends whose parents had already departed. Ella came over and sat down.

When the story about the father hare and his son was over, Hannah and Ella wandered off to the sand table.

A mother dressed in a business suit came in with her son in tow. "Hi, Sarah," said Kelly, and she couldn't help comparing the woman's perfectly tailored suit with the drab maternity dress from Target that she had hurriedly donned before hustling Hannah and ten-year-old Grace out the door and into the car. Between dropping Grace in front of the neighborhood elementary school and heading downtown, Kelly had managed to apply lipstick using the visor mirror, but her hair was still just hastily pulled back in a tight elastic.

"Hi, Kelly," Sarah said, glancing at Kelly's abdomen. "You look great." Then she turned to her son and kissed him. "'Bye, Sammy. Mommy's got to run." As she extricated herself from her son's embrace, she said to Kelly, "God, I hate always rushing off." And then she waved a carefully manicured hand at Kelly with the apologetic look of a working mother on a tight schedule.

"I know what you mean," Kelly started to say, but Sarah was already too far away to hear.

Kelly stayed and watched the kids for a few more minutes; when the children were called to circle time, she quietly slipped out of the classroom, made a quick stop at the supermarket, and headed to her car.

The next stop was across town at Brigham and Women's Hospital; she had a long prenatal appointment during which she would have a blood test for gestational diabetes, a common condition she had had during her pregnancy with Hannah. Kelly was twenty-six weeks along in her pregnancy, and she was pretty sure she'd fail the test and be required to start the restrictive diabetic diet to control her blood-sugar levels and avoid insulin injections. She smiled as she recalled her carbohydrate binge of the previous day, which she'd relished in the full knowledge that it might be her last for several months.

Kelly pulled into the parking lot, checked in at the clinic, and had her first blood sample drawn. Then she drank the super-sweet cola that challenged her body's ability to metabolize sugar and sat down to wait exactly one hour until the next blood sample was taken. She pulled a manila folder out of her bag and began to read through the information she'd printed out from the website of Boston Children's Hospital, where she would interview the following week.

Before leaving Baltimore to move to Boston, Kelly had completed the coursework for a doctorate in psychology. As a student, she had excelled at working with children who suffered from chronic diseases such as kidney failure, cerebral palsy, and severe developmental delay. She was pleased when the best children's hospital in Baltimore had aggressively recruited her for her required one-year internship, the last thing that stood between her and her degree. But instead of taking that coveted position, Kelly had put her own career on hold and moved to Boston—so I, her former-journalist husband who had just graduated from medical school, could do my residency in obstetrics and gynecology at Harvard-affiliated Brigham and Women's Hospital.

Our plan was working perfectly—Kelly was helping our girls adjust to a new city and was pregnant with our third child during this year after our move to Boston and before her anticipated return to work. She had applied widely to internship programs in Boston and found that her success in Baltimore was opening doors in her new hometown. Leaders of the top programs in the city offered her interviews, and when the calls from internship directors eager to have her join them followed her visits to their hospitals, Kelly realized she would have her pick of jobs.

A young phlebotomist with a bored look on her face called Kelly in for her second blood draw. This one hurt, as the woman missed the vein on the first jab and had to reposition the needle. Kelly went back to her reading and waited for the third blood draw, an hour later.

At some point during the next hour, it occurred to her that she had not felt the baby move for some time. As the minutes ticked by, Kelly started to become anxious. Then there was a kick and a twitch as the baby shifted position. Kelly picked up her reading again, but her anxiety didn't dissipate entirely. Perhaps it was the fact that she had been required to fast since the previous evening because of the blood test. After the third blood draw, she paged me.

"Hey, what are you doing?" she said, trying to sound relaxed.

When she called, I'd just finished seeing a patient in the emergency room two floors below where Kelly was, so I was able to take a break.

"I'm just downstairs. I'll be up in a minute."

I found Kelly sitting forward in her chair trying to concentrate on the sheaf of papers in her hand.

"I'm sure it's nothing," she said. "It just doesn't feel right."

"What do you mean? Is the baby moving?"

"Yes, she just moved, but something doesn't feel right." She didn't look all right.

"I'm sure it's fine," I said, still in my practiced reassuring-physician mode. "Let's go have a look." I took Kelly's hand and led her through the reception area of the obstetrics clinic, past a few women who were waiting to be seen, and found an exam room that was unoccupied. Kelly climbed onto the table while I retrieved one of the portable ultrasound machines from the hallway. Then together, we looked at our baby happily ensconced in Kelly's protuberant belly.

Larissa was moving intermittently, reaching out to touch the uterine wall, then touching her tiny face with fingers that were each perfectly formed but no larger than a piece of macaroni. As we watched, she brought her legs up and then kicked, which made Kelly smile as she felt the movement we could both see on the screen. We gazed for a few minutes at our oblivious child suspended in amniotic fluid, watching her reach and turn and make the other normal movements of a developing fetus that indicate a healthy pregnancy.

Kelly smiled and looked up at me. "I feel better." The inexplicable stress that had been building since she arrived at Brigham and Women's began to dissipate. She had her last blood draw and headed straight for the cafeteria. A few minutes later she was on her way back across town to the nursery school, eating a sandwich and listening to a National Public Radio call-in show featuring rescuers of a recent disaster. Yet somehow, inside her SUV with the heat on full blast, Kelly felt insulated from what was still raw.

JANUARY 11

Early the following morning, exactly three months and two days before her April due date, Kelly was awakened by an uncomfortable sensation. It began like dull menstrual cramps, and then her lower abdomen tightened to the point of pain before slowly relaxing, leaving a hollow ache behind.

This happened repeatedly, and for a long time, Kelly lay in the dark waiting for the cramps to go away, certain that this unusual sensation was just another in the catalog of pregnancy oddities. When my alarm rang, and I stumbled to the shower, she feigned sleep.

Contractions before the due date are not necessarily unusual; in fact, isolated contractions are common in the last months of pregnancy as the uterus prepares for labor by tightening and relaxing the muscles that will be called on to push the baby through the birth canal. They generally dissipate quickly. But these contractions increased in strength to the point that Kelly was becoming uncomfortable. She struggled to stay calm as her abdomen kept tightening and relaxing, every six minutes, then every five minutes. Kelly had been in labor before, and these cramps felt more and more like labor and less like the occasional twinges of the third term. She called me out of the bathroom, where I was getting dressed for work.

"I'm having contractions," Kelly said, with urgency bordering on panic in her voice. She scrunched her face up, and I wasn't sure if she was in pain or just trying to keep the tears of disbelief from escaping.

"I'm sure it's nothing," I said, certain of my words. "They are probably Braxton Hicks," I reasoned. This is the term for isolated contractions of late pregnancy that are not worrisome, named for the physician, John Braxton Hicks, who'd first described the benign phenomenon.

Nonetheless, I was concerned enough to call my parents, who lived in the next town, and ask them to come over quickly—we needed them to watch Hannah and Grace, who were asleep in their rooms. I wanted to get to the hospital as soon as possible so Kelly could be evaluated but also so that I wouldn't be too late for work. I wasn't terribly worried—I put on a tie while we waited for my parents so that I would be ready to start my day at the hospital once Kelly was sent home.

Twenty minutes later, my parents, still in their pajamas, arrived on our doorstep. Kelly was already in the car. I fumbled for my keys and dropped them; I bent down and felt the cold concrete as I reached across the dark driveway.

As Friday dawned, we drove through the winter streets of Boston, which were narrowed by the snow pushed to both sides by the plows. I drew up to the main doors and left the car outside the hospital with two sleepy valet attendants. Inside, a slightly more alert volunteer took us to Labor and Delivery.

Though it was barely dawn, the unit was humming. Machines beeped; the television attached to wall in the waiting room droned; a pair of soon-to-be grandparents dozed, leaning against each other; and harried nurses, getting ready for their change of shift, exchanged information. A slim nurse in her fifties escorted us to a room with a sliding glass door and expertly hooked Kelly up to two monitors.

One sent ultrasound waves into Kelly's uterus so we could hear the flow of blood through the fetal heart. This heart rate was plotted on a slowly moving strip of paper and it scrawled out the expected jagged line, reassuring us that our baby was healthy in the protective confines of the womb.

The other monitor contained a pressure sensor that measured uterine contractions; this information was also plotted on the strip of paper. I was looking for the flat line of uterine rest, which would tell us that the contractions were mild or absent. But instead, there were smooth bell-shaped hills coming regularly, every three minutes, again and again and again, plotting a pattern of contractions that looked more and more like labor.

In the vast majority of pregnancies, a perfectly timed cascade of hormonal signals between the developing fetus and the mother initiates labor after approximately forty weeks of pregnancy. Contractions develop, the cervix dilates, and the outcome is a well-developed healthy newborn.

True labor can also be triggered prematurely. Sometimes, labor is precipitated by chemicals released when the cervix gives way and dilates preemptively. Other times, the hormonal cascade is initiated by a stressed-out fetus that is unable to grow adequately within the womb. If the labor is caused by cervical failure or fetal stress, the best solution is to stop the contractions and fix the problem—attempt to reinforce the cervix or mitigate the fetal stressors.

Labor can be caused by more insidious factors as well, such as bleeding in the space between the placenta and the uterus, or an infection that breaches the membranes that protect the fetus and invades it and the surrounding amniotic fluid. In these cases, labor is an adaptive response, a way for the mother to get rid of the pregnancy before she bleeds to death or is overwhelmed by infection.

However, when a woman shows up on Labor and Delivery complaining of contractions far in advance of her due date, it is almost never clear why the patient is contracting. Aware of the potentially devastating consequences for the fetus of an early delivery, most obstetricians will try to stop preterm labor until they have evidence that a force of nature stronger than their medicine is at work.

Ultimately, it is the cervix—the conical structure that extends down from the uterus into the vagina—that is the arbiter of labor.

Typically fibrous and strong, the cervix serves to protect the fetus from the outside world for nine months. At the very end of pregnancy, the composition of the cervix changes and it becomes receptive to the contractions that will cause it to dilate and shorten from a long and impassable canal to the ten-centimeter-wide aperture that's big enough for a baby to pass through. Until the cervix shortens and dilates, contractions are essentially meaningless. *Surely,* I thought, *these are Braxton Hicks contractions. It is only a matter of time before they resolve.*

A senior resident came to examine Kelly. He was somewhat disheveled, and I knew he'd been up all night. Gently he placed a speculum in Kelly's vagina so he could see her cervix. With the nurse angling a floor lamp over his shoulder, the young doctor adjusted the speculum to make sure of what he was seeing.

"She's three and a hundred," he finally said to me, using the shorthand of our field to describe Kelly's cervix as three centimeters dilated and completely shortened. He looked scared, without the confidence he would have shown if Kelly hadn't been a colleague's wife.

"What does that mean?" Kelly beseeched. The resident looked at me, unsure of how to respond. After a moment, he said to her, "You might be in labor." I opened my mouth to speak, but no words came out. "I'm going to talk to the nurses," he blurted on his way out.

I began to feel the stirrings of a dread so foreign I had no name for it. Carefully, I willed the corners of my mouth to rise into a smile and masked my fear with a loving and reassuring look that belied the alarm bells going off in my head. These were not Braxton Hicks contractions.

The exam galvanized the nurse taking care of Kelly. She efficiently recruited other nurses, and they formed an expert and well-practiced team. One brought an IV pole to hang a bag that contained magnesium in the hope that this medication would cause Kelly's labor to abate. Another rolled Kelly onto her side and injected steroids into her buttock to help our baby's lungs mature more rapidly. A third nurse, a matronly type, hovered, asking how she could help; the nurses were used to teaming up when a patient in preterm labor rolled through the door, and they were particularly attentive when the patient was one of their own.

I watched the magnesium drip from the IV bag into Kelly's arm as we kept vigil by the contraction monitor. I counted the minutes between the smooth, round hills and wished that the intervals would increase, signifying a slowing in her contraction pattern.

"Maybe the magnesium will stop them."

"Maybe." Kelly winced as another contraction began.

Obstetricians count weeks of pregnancy (the fetus's gestational age) starting from the first day of the patient's last period; the due date is set as forty weeks after that. Any baby born before thirty-seven weeks is considered preterm. Until that point, the fetus is not entirely ready to enter the world, and she suffers the consequences of prematurity when she arrives too early. The earlier the baby comes, the greater the risks.

In the United States, approximately 13 percent of all pregnant women deliver early, usually only a week or two. Only 2 percent of pregnancies end before thirty-two weeks. Kelly was twenty-six weeks into her pregnancy when she arrived at Brigham and Women's in labor. And as rare as it is, preterm delivery is practically unheard-of in a woman who has carried two children to full term. For Kelly, who had delivered Grace and Hannah within days of their due dates, a preterm delivery seemed an impossible nightmare.

I could see fear on Kelly's face—fear of what would happen if our daughter was born prematurely, but mostly fear of what she didn't know. The fear was interrupted, at predictable intervals, by her uterus, which caused enough pain to wrest her focus away from what might be coming and back to the present situation. In the years since that morning, I have cared for hundreds of women in preterm labor, and I have seen that same look on the faces of my patients and their partners. I always try to reassure them—not that everything will be all right, because I don't have that kind of influence, but that I will do everything in my power for them and for their child.

As I sat next to Kelly, my tie began to constrict my throat, and a knot formed in my stomach. I knew specifically what to worry about, and, unlike Kelly, I was not distracted by pain. The terror began to take hold, and I felt the cold run down my torso—the cold reality of Larissa's unpreparedness to exist outside of Kelly's womb, and our unpreparedness to deal with that.

Had Kelly's doctors tried to reassure us that morning—which they didn't; perhaps they knew better than to try—I would not have been reassured, because I knew the awful uncertainty about delivery twenty-six weeks into a pregnancy. Over and over I thought, *This doesn't make any sense.* I kept waiting for the contractions to space out and then stop, but the black pen in the contraction monitor continued drawing hills. I wanted to shut off the power to the monitor to make the pen stop drawing hills. It was like Kelly and I were in a car that was about to crash, but for some reason I couldn't stomp on the brakes.

For an hour, the magnesium dripped in, but hills on the monitor kept coming. Then the contractions became stronger.

The attending obstetrician came in. She looked at the contraction pattern on the monitor, and then spoke to Kelly. "Would you like to have an epidural? If the contractions space out, we can always take the catheter out of your back." She didn't sound like she was optimistic the epidural would be removed anytime soon.

Before Kelly could respond, a gush of fluid poured out between her legs and soaked the linens. Like a dam giving way, Kelly's water had broken. In that instant, I knew that my daughter would be here soon. Kelly's cervix was now dilated to six centimeters. Delivery was inevitable.

An obstetrician has two patients: the mother and her fetus. When the mother is otherwise healthy but is in preterm labor, the baby becomes the focus of some rather grim calculations.

Babies that are preterm but born after thirty-two weeks of pregnancy tend to do very well. Occasionally, those children will suffer mild behavioral or learning disabilities that become apparent in elementary or middle school, but the risk of serious complications is relatively low, and almost every baby survives.

Babies born before twenty-four weeks of pregnancy—the point when a baby weighs about one pound—die more often than they survive. Those babies that do live suffer significant complications, including severe lung injury, cerebral palsy, mental retardation, injury to their intestines, and a host of other, less severe problems. Although doctors will occasionally and reluctantly try to save a twenty-two- or twenty-three-weeker, most of the time a baby born before twenty-four weeks is gently wrapped in a blanket and allowed to die in the parents' arms.

Of those babies born between twenty-four and thirty-two weeks of pregnancy, over 80 percent will survive, and with each passing week in the womb, a baby becomes less likely to suffer major complications related to prematurity. Had Kelly gone into labor three weeks earlier, there would have been no hope. We would have waited for her to deliver, and then mourned the inevitable death of our child. If Kelly had gone into labor three weeks later, we would have been cautiously optimistic about our baby's chances of a speedy recovery and a normal infancy.

After the epidural was in place, Kelly's pain abated, and she was able to focus entirely on the imminent delivery of our premature baby.

"Will she live?"

"Probably. Survival is eighty percent."

"Are you worried?"

"Yes."

"What about?"

"There may be complications."

"I'm so sorry." Kelly started to cry. "I'm so sorry."

"Oh, sweetie, it's not your fault." I was up on the bed beside her now. "It's not anyone's fault." As Kelly's tears fell, I held mine back. I felt like I needed to remain in control. As if somehow, if I could force back my fear, I could prevent this awful event.

Awful event? This was the *birth of my child.* A precious little girl who would be the third sister. The one who was supposed to be the minx, the youngest daughter who is so adorable and doted upon that she gets away with murder. How could this be happening?

Then the tracing of Larissa's heart rate, which had zigged and zagged in a reassuring jagged line at around 150 beats per minute, showed a dip as one of Kelly's contractions temporarily squeezed off the supply of blood. The heart rate normalized when the contraction abated, but a few contractions later, we saw another dip of the heart rate. This pattern suggested that several hours into Kelly's labor, our baby was beginning to get tired, and she might not tolerate the remaining contractions necessary for her to reach delivery.

The chief resident came into Kelly's room, accompanied by the attending obstetrician.

"We have, um, been watching the, uh, fetal heart rate," the resident began haltingly, glancing at his supervisor periodically for support, unsure of whether to address Kelly or me.

His supervisor came to his rescue. "I think the question is whether your baby will tolerate labor."

Tiny and ill-prepared to enter the world, Larissa might not live through the delivery process. We decided that her chances for survival would be better if Kelly had a cesarean section.

I left the labor room. A few steps away, a group of my fellow residents were standing together in the hallway; they looked up at me, and then back at one another, unsure of how to respond to their colleague whose wife had become a patient. One of them left the group and came over to give me a hug. "It will be all right," she said.

"Thanks," I managed, not believing her, and I pulled away to avoid letting her see the tears well up.

I went upstairs to the locker room and changed into standard-issue blue operative attire as I had dozens of times before scrubbing into similar surgeries. A gray-haired obstetrician, unaware of why I was changing, made small talk.

"Quite a winter we're having."

"Sure is," I answered unenthusiastically.

"You got a case?"

"C-section," I answered, then hurried out of the locker room before he could ask me for any details.

"Have a good case," he called after me.

Kelly's bed was unplugged, the connections to the monitors that had tracked the progress of her labor were removed from her abdomen, her IV bag and her epidural syringe were disconnected from the wall, and she was rolled down the hall past a half a dozen delivery rooms to the operating room, where under the bright lights, she was transferred to the table.

The attending obstetrician and a senior resident were gowned in blue, their faces mostly obscured by masks. The laser intensity of their eyes conveyed what they didn't say: This might not end well. Their conversation was sparse.

"Curved Mayos, please."

"Kocher's."

"Metzenbaum scissors."

The room was otherwise quiet.

As they cut through the layers of the abdomen that separate the uterus from the skin, they encountered scar tissue left over from Kelly's two prior C-sections. Scar tissue also encased the uterus, making the going slow. Finally a vertical incision opened Kelly's uterus like the first cut into a pear. But our daughter's head was wedged down into the pelvis, caught in the scar tissue, tight bony pelvic structures, and the dilated cervix. Watching over the blue curtain placed to spare my wife a view of her viscera, I saw the placenta deliver itself through the incision. Without the placenta attached firmly to the uterine wall, our child would have no source of oxygen until she too was delivered and could breathe on her own. Meanwhile, the unattached placenta became a vent for Larissa's own blood, which began to hemorrhage onto the operating field.

"What's going on?" Kelly asked. She could see the horrified expression on my face.

"They are trying to get her out."

"Why is it taking so long?" Kelly asked with genuine panic in her voice.

"She's stuck."

"Um, I can't get under the head," the resident said, looking up briefly to address the attending obstetrician. The resident removed his hand, and the more senior doctor tried to gently insert her hand below Larissa's head in order to bring it out of the pelvis.

A minute went by.

"Forceps," the surgeon called, and a nurse rushed out of the room and returned seconds later, unwrapping the sterile packaging around the long steel instrument as she came through the door.

First the resident and then the attending failed to get the forceps past Larissa's head to the point where they could free her.

"Could I have a hand from below?" When the attending called for this, her tone was higher. I recognized this elevated octave as desperation. Doctors feel it too. When lives are on the line and outcomes uncertain, the pressure to perform miracles is daunting.

Knowledgeable enough to understand that a disaster was unfolding but lacking the skills or standing to do anything about it, I buried my head next to Kelly's, unable to watch. Kelly reached up with the one arm that wasn't strapped to the table armrest and held on to me.

A semicircle of nurses had formed around the operating room table, waiting, barely breathing, for the surgeons to free Larissa. Now, one of them darted forward, reached under the sterile blue drapes into Kelly's vagina, and pushed Larissa's head up into Kelly's pelvis.

Hours later—really just three or four minutes after the placenta delivered through the uterine incision—the chief obstetrician lifted our little girl out of her mother's pelvis.

The umbilical cord was briskly clamped in two places and cut, and the obstetrician rushed Larissa over to a corner of the operating room where a team of neonatologists stood huddled around their own equipment. The obstetrician laid Larissa down on her back with her head positioned toward the most doctor-accessible edge of the baby warming table, and she stepped away. The team of specialized pediatricians converged on my daughter.

There is some evidence that the risk of brain hemorrhage in premature newborns can be reduced by minimizing the shifts in pressure that the baby is subjected to. With this in mind, I had requested that the senior neonatologist personally perform the intubation—the insertion of a breathing tube down Larissa's trachea—instead of allowing a resident to perform this delicate task.

"Breath, breath, breath . . ." The attending neonatologist counted them out. "Watch the pressure." She spoke with authority to the resident at her side while she held a tiny mask over Larissa's nose and mouth.

"What's the pulse?"

"It's good. A hundred and fifty," responded a nurse who had gently pinched the bit of umbilical cord going from Larissa's abdomen to the clamp so she could feel the pulse in an umbilical artery.

"Would you like help with the intubation?" a resident asked her attending, euphemistically asking if she could perform the procedure.

"No. I got it," the attending responded curtly.

Several seconds later she put aside the mask, reached for the curved stainless steel laryngoscope, and, in a single practiced maneuver, slid it down to the base of Larissa's tongue and lifted the instrument up to expose her vocal cords, which formed the top two sides of a narrow triangle that framed the view down Larissa's trachea.

"Tube," she called, and a tiny breathing tube was placed into her outstretched hand. She slid the tube between Larissa's vocal cords.

"Breath," she commanded seconds later. She exchanged the laryngoscope for her stethoscope and confirmed that the tube was correctly situated by listening for the sound of air moving in each lung. Satisfied, she taped the tube in place. Within sixty seconds of being born, Larissa was being dried off and wrapped in warm blankets.

By the time she was five minutes old, the neonatologists had packaged her into a Plexiglas incubator and were rushing her out the door of the operating room. Before they left, they rolled her to where Kelly could see our little girl, wrapped up in the box with the tube down her throat connected to the enormous purple bulb that one of the doctors kept squeezing. "Hi, Mom," one of them said in a falsetto voice, and then they rushed her off to the newborn intensive care unit to stabilize her. Their hurry to get back to the unit did not comfort me.

CHAPTER 2

Critical Hours

First, the air had been punctuated by the staccato of surgeons requesting instruments and ordering maneuvers while they struggled to free our child from Kelly's abdomen. Then, the chatter of the neonatologists coordinating their resuscitation efforts filled the space. Now a hush fell on the operating room. Grim in the wake of an imperfect delivery, the obstetricians methodically set about restoring Kelly's anatomy, layer by layer, in a practiced routine they had done hundreds of times.

"Zero Monocryl," one said softly, unhurriedly asking for the deep blue suture to close the uterus.

"How about something to take the edge off?" the anesthesiologist asked. He had already drawn up a syringe of the antianxiety medication Ativan, and he injected it into Kelly's IV. She was soon asleep under the kind, sedating influence of the drug.

I sat quietly by her side, holding—really gripping—her hand. In front of me, the blue curtain rose, protecting the sterility of the operative field and shielding the bloody surgical space from Kelly's view. I knew what was going on behind the blue curtain; the surgical repair was straightforward, even mundane. I also knew that one floor above us, Larissa was trying to survive. I could imagine what was going on up there too. It wasn't mundane.

"Give us a couple of hours to get her settled," one of the nurses had said as the team of pediatricians left the OR. "If anything unexpected happens, I promise I'll come and find you."

What she meant was, If Larissa's going to die, we'll come tell you.

Outside the operating room, the stocky nurse pushing the wheeled incubator took charge, like an engineer driving through a rail yard. "Excuse us," she barked loudly to anyone who even threatened to get in the way. The large purple breathing bulb made a *whishhh* sound when the respiratory therapist squeezed it and a droning *vzzzzz* sound when he released it.

The nurse reached for the keycard hung on her ID necklace and jammed it into the slot next to the elevator call button. The door opened within seconds as the key overrode the elevator computer system and summoned the nearest car. "Please step out and wait for the next one," she commanded to the people in the elevator. An elderly couple on their way to visit their newest grandson hurried to comply.

The team rode up, and when they got off the elevator, the nurse drove straight to the new-arrival section of the NICU where a station had been hurriedly prepared for Larissa. Nearby, tiny babies lay surrounded by large machines that provided oxygen, intravenous nutrition, antibiotics, fluid—whatever was needed to sustain life. Here, under bright lights, cared for by an army of nurses and doctors, babies who need help spend their first hours, weeks, and sometimes months.

The roof to Larissa's incubator was swung up, and, in a practiced motion, six hands placed Larissa, her breathing tube, and the spaghetti of monitoring wires onto the table where the neonatologists would do their best to re-create Kelly's womb for our daughter.

The table, designed for maximum access and positioned at a comfortable height for standing physicians and nurses, was surrounded on three sides by low Plexiglas walls that were just high enough to prevent Larissa and the syringes, suction bulb, and assorted equipment from rolling off. The back of the table rose up to support a heat lamp, which glowed orange overhead. Whenever Larissa's temperature fell below 98.6 degrees Fahrenheit, the probe that was stuck to her chest would signal the warmer to increase the radiance of the lamp. To one side, ribbed plastic tubing extended from a square, beige, wheeled ventilator and snaked onto the table. Carefully holding on to Larissa's breathing tube, the respiratory therapist disconnected the plastic purple bulb and attached her tube to the ventilator tubing. The therapist weighed at least three hundred pounds and was built like a linebacker, and his hands were bigger than Larissa's entire body; it was nothing short of amazing that this man's beefy fingers were able to connect the delicate tubes that kept the tiniest of babies alive. With everything in place, he hit a switch to turn on the ventilator and checked the settings carefully. Then he easily moved on to the next sequence of tasks, his heavy breaths mingled with the soft but sure whooshing of the ventilator as it pumped the oxygen that Larissa needed to survive.

With the orange heat lamp shining down on her, Larissa lay naked but for a doll-size diaper as she was connected to an increasing number of tubes and wires by adult hands moving in and out from the sides of the table. One of the doctors cut another few millimeters off Larissa's umbilical cord, exposing the vein and two arteries, and threaded tiny catheters down these easily accessible vessels. Normally, these blood vessels, which are a fetus's lifeline in the uterus, clot off and are never used again once the umbilical cord is clamped and cut immediately after delivery. But in the case of a very premature baby, doctors use the umbilical arteries to measure blood pressure, and the umbilical vein to infuse fluids and medications.

Another doctor put a tiny flashlight behind Larissa's arm and used the light to identify a vein; it made a dark line against the illuminated red-yellow of her nearly translucent arm. Guided by this backlight, he managed to stick a needle into this tiny vessel, and then he threaded a catheter up inside her arm so that Larissa could receive fluid and liquid nutrition.

The respiratory therapist, relieved of his job of squeezing the purple canister, briefly disconnected the breathing tube from the ventilator and squirted a type of fluid called surfactant down the tube; this would help keep her lungs from collapsing. One of the less experienced nurses, still tentative in her motions, was allowed to place electrodes on Larissa's chest and attach them to machines that measured her heart rate and respiratory rate. She also carefully taped a tiny red photodiode to Larissa's foot; this would measure the oxygen level in her blood. Once all the lines, tubes, and machines were attached, two hands elevated Larissa while two more slid her into a plastic bag up to her neck; this was to limit fluid loss through her translucent skin.

Two hours later, I left Kelly in the surgical recovery room and walked upstairs to the NICU. I showed my hospital identification and was waved in.

"Baby Lowery?" I asked an orderly in the hallway, referring to Kelly's last name, which would be Larissa's until discharge.

She pointed across the unit. I walked through the crowded room, dodging ventilators, medication carts, and laundry bins. I passed a dozen babies lined up on both sides of the room, each one surrounded by equipment.

These were the sickest children, mostly confined to their raised tables or incubators. A mother sat in a rocking chair, pushed right up next to one of the incubators. She wore the loose-fitting sweatshirt common among women recently postpartum, whose abdomens still protrude. Her hands and her arms disappeared up to her elbows inside the incubator through the two portholes in the side of the device, and with the index finger of one hand she stroked her son's leg. Her face was pressed to the Plexiglas of the incubator, and I could make out only the occasional soft word she spoke through the glass: "You can do it, Jason. . . . Mommy loves you so much. . . . What a good boy."

At the farthest end of the room, portable blue barriers created a makeshift semblance of privacy around one of the incubators and a couple of rocking chairs. A hush seemed to surround this part of the otherwise busy NICU, and nurses and staff members were noticeably quieter when they cared for the babies near those partitions. While I made my way across the room, a nurse respectfully stuck her head around the barrier, exchanged a couple of words with another nurse, and then withdrew. As I got closer, I could hear faint sobbing coming from behind the screens. Soon a black-clad hospital chaplain emerged, head bowed after giving last rites, and walked purposefully toward the exit. This was a place where birth and death often came close together.

I found Baby Lowery ensconced in her plastic bag. I knew it was her only because a hastily scrawled sticker with her name in black marker had been stuck to the wall behind her head. My tiny daughter weighed one pound, fifteen and a half ounces, about as much as a quart of milk. She made her unhappiness at being born quite clear: her arms and legs, so scrawny without baby fat, gesticulated wildly as she searched for the confines of her mother's womb. Everything dwarfed her—the IVs, the plastic tubing hooked to her breathing tube, the syringe lying nearby that was longer than her leg.

I must have stared for some time. I had seen many premature newborns, but it took me a while to match the image of this tiny, thin, plastic-encased creature with my concept of a daughter.

A young nurse came up. "Can I help you?" she asked.

"I'm Dr. Wolfberg . . . Adam . . . I'm the father."

"Oh." She didn't bother hiding her frown as she looked at Larissa, then at my scrubs and my hospital ID, and finally at my face, as if she were trying to reconcile discordant concepts. "Hang on. Let me get her doctor," she said, and she fled to retrieve the neonatologist in charge of Larissa's care. I continued to stare, standing tentatively a foot away from the table.

Hi, Larissa. I formed the words with my mouth, but they didn't come out. "Don't worry, sweetie. They're going to take really good care of you here." I was whispering, willing the words to penetrate her tiny being and offer some comfort. "Mommy and I love you."

"I'm Dr. Abdulhayoglu." The petite physician with her brown hair pulled back in a short, efficient ponytail introduced herself. Her last name was so difficult for Americans to pronounce that everyone called her Dr. Elisa, mixing the respectful title with her first name. Given her height and youth, the name fit.

"How's she doing?" I asked.

"It's really too early to tell." Dr. Elisa avoided the question, but she smiled up at me, acknowledging the dodge with compassion.

I was still a short distance away from Larissa. I could have touched her, or at least touched her bag, but I didn't.

"The first forty-eight hours are critical," Dr. Elisa started. "If she makes it to Monday without anything major happening"—it was Friday afternoon—"I'll be very optimistic."

"What are the chances of that?"

"It's really hard to say." Dr. Elisa avoided this question also. "We'll take it day by day."

"What will you be watching for?" I pressed. I remembered the basics of what could go wrong early in a premature baby's life, but I wanted to hear it again, and I wanted to hear whether anything in Larissa's first hours had raised a red flag for her doctor.

Dr. Elisa did her best to lay out the issues that Larissa would face in the short-term and explain the technology that was available to help her survive.

"During her first few days she could start to breathe for herself, or her lungs might begin to fail," the neonatologist cautioned.

Indeed, over 40 percent of babies who weigh less than three pounds at birth have some degree of breathing impairment, and more than one out of every five babies in this group will have long-term breathing problems. Overall, lung disease is the leading killer of the smallest babies.

Dr. Elisa continued, "The first days after delivery are also the most likely time for an IVH to occur." I knew this was medical shorthand for intraventricular hemorrhage, bleeding in the brain, which is particularly common among premature newborns and is the injury most often responsible for lifelong disability.

She concluded with a sober reminder. "Since—as you know—a common cause of preterm labor is an infection inside the uterus, we'll be on the lookout for signs that Larissa is going into shock due to an overwhelming infection." I remembered that up to 20 percent of extremely premature newborns develop systemic infections and that this increases the risks of other complications.

Dr. Elisa and I stood next to each other clad in identical blue scrubs, our arms crossed in front of ourselves, as if we were warding off the cold.

Listening to her talk, I saw Larissa in front of us, but my mind kept leaping to the pictures of children I had seen in medical school and residency. The child in a wheelchair, muscles contracted by cerebral palsy, who had suffered a severe intraventricular hemorrhage at birth; the child moored to an oxygen tank by thin plastic tubing.

She smiled. "You can touch her, you know."

I edged closer to the table and reached out my hand. I touched the plastic bag, and then, with some effort, I found a patch of skin by her temple that was not covered with tape, a hat, or the bag, and I gently caressed my daughter for the first time. I felt a rush of conflicting emotions—a desire to do anything in the world that would make Larissa okay and an almost panicky wish to follow the chaplain out the door; a sense of overwhelming tenderness for my unbelievably frail daughter, and a feeling of revulsion for this scrawny creature who was red and purple, not pink, and pinned down by wires and tubes, not a cozy receiving blanket.

"You know what you can do?" Dr. Elisa said. "Your wife's blood doesn't match Larissa's, which means yours does. Why don't you go bank some blood, because Larissa will inevitably need a transfusion at some point in her hospitalization."

This suggestion—and given the safety of the hospital blood supply, we both knew it was just her attempt to help me feel useful in this situation that was entirely out of my control—gave me a purpose. I walked the length of the hospital to the blood bank and eventually reclined in the vinyl chair as the technician swabbed my forearm with alcohol and easily, without searching, stuck a large catheter into a vein. As the dark blood ran down the tubing into a collection bag, I closed my eyes, trapping the tears before they fell.

I found Kelly in a postpartum room upstairs from the NICU, where, exhausted by the morning, she was trying to take a nap. I left the hospital to pick up Grace from school. Earlier, Hannah, three years old and oblivious, had happily gone back to our house with my parents.

I parked behind the school and found Grace working on her homework in the after-school classroom. Grace, who had the palest blond hair and a moon-shaped face, was laughing with a group of friends who were working on math problems together at a table. She had always been a self-assured child and had quickly made friends when she started fourth grade in Boston. She came over to me and we collected her coat, hat, and mittens, put her lunchbox and homework in her backpack, and walked in the twilight to the car.

Inside the car, on the short drive to our house, I said to Grace, "Your mom had the baby this morning. You have a sister."

"I thought Mom was going to have the baby in April."

"She was, but the baby came early."

"Oh. When will she come home from the hospital?"

"Well, she's really, really small, Gracie."

"No, not the baby—Mommy. When will Mommy come home?"

"Mommy will come home in a couple of days. But your sister is very small, and she's pretty sick. She will have to stay at the hospital for a while."

"Is she going to die?"

"I'm not sure," I said.

We had barely walked into the house when the doorbell rang. I had called Kelly's sister from the hospital that morning; Karen had dropped her kids at school in Montgomery, Alabama, and driven straight to the airport. Now, she came in the door and swept Hannah and Grace up in an embrace.

"Why are you here, Aunt Karen?" Hannah wanted to know.

"To see you," came the reply. It never occurred to Hannah not to believe her.

Later, while Hannah was eating dinner, Grace pulled Karen aside. "Did you hear that my mom had my sister?"

"I did, sweetie."

"I think she might die."

"I hope not."

"Me too."

There was a pause.

"My mom's going to be okay, right, Aunt Karen?"

"Your mom is going to be fine."

"You sure?"

"Promise."

That night, I helped Kelly move from her hospital bed to a wheelchair. She held a folded blanket to her abdomen to minimize the pain as I moved her legs over the side of the bed. They were puffy from the combined effects of pregnancy and surgery. After she was draped in blankets against the chill of the hospital, I wheeled my wife to the NICU. We paused at the front desk, and one of the nurses came to retrieve us. She wheeled Kelly slowly to Larissa's bedside. I followed a few steps behind.

The nurse touched a lever, and the Plexiglas-circled table dropped down to Kelly's level. Tentatively, my wife extended her arms toward Larissa, grimacing with pain, and rested them on the Plexiglas. She stopped and turned to the nurse.

"Can I touch her?"

"Of course you can."

Kelly put out her hand just as Larissa thrashed her arms and legs against the plastic. Startled, Kelly drew back.

"Sometimes, you can settle them by gently cupping them with your hands," the nurse explained, reaching in and taking Kelly's now tentative hands.

One of Kelly's pale hands cradled Larissa from underneath, the tips of Kelly's fingers coming to rest against the dark fine hair on the back of Larissa's head. Kelly's other hand gently bent Larissa's legs through the plastic, enclosing them, limiting them, as her womb had done until that morning. Larissa kicked again, and this time Kelly's hands remained gently in place. Larissa seemed to calm at the touch.

"I am so sorry," Kelly whispered. "I am so sorry." Tears began to run down her cheeks.

I was crying now also, rubbing Kelly's back.

Larissa, stilled by her mother's warm hands, lay quietly, her thin chest rising and falling with the whoosh of the ventilator beside the table. When the ventilator exhaled, it created a divot in the skin just below her sternum, making it look as if Larissa's chest were caving in.

With a small pop, a teardrop hit the plastic. Kelly wiped her eyes with her shoulder so as not to contaminate Larissa's environment.

On rounds the next morning, the pod of people moved awkwardly from bed to bed, stepping around the equipment and chairs that crowded the NICU. The Brigham NICU is an enormous room that fills most of the sixth floor of the building. Half walls, each rising five feet, divide the room into fourths—NICUs A, B, C, and D—and babies are lined up along these walls and around the periphery of the unit. The two units in the center, B and C, are for the sickest babies, and because of the configuration, it's a noisy place without adequate storage space for all of the equipment needed to keep the babies alive. The fellow—the most senior physician in training—nominally led the rounds and served as the facilitator for the group of residents, medical students, nurses, and respiratory therapists. But when an important decision needed to be made, all eyes turned to the neonatologist in charge; that day, it was Dr. Linda Van Marter.

Reaching Larissa's bedside, the members of the group arrayed themselves in a semicircle around Larissa and spent extra time there because she was a new admission.

"This is day of life two for Baby Girl Lowery, born yesterday by cesarean to a thirty-eight-year-old wife of an obstetrics resident here at the Brigham," intoned the resident assigned to take care of my daughter. There was an audible gasp from several members of the group at the news that this scrawny child belonged to a member of the hospital family.

The resident continued, "This baby's hospital course has been relatively unremarkable. She was lined and labbed, pan-cultured and started on triples"—medical shorthand to describe routine procedures: intravenous lines placed, blood and urine sent to the lab for the usual measurements and to test for infection, and broad-spectrum antibiotics administered to treat any combination of bacteria trying to sneak up on Larissa.

The resident read the ventilator settings off her rounding sheet, and in a workmanlike monotone listed the contents of the intravenous nutrition that dripped slowly into the catheter in Larissa's arm. "Her vital signs are within normal limits, she is requiring no medications to maintain her blood pressure, and her morning labs today are all normal except that her hematocrit is a bit low, presumably an equilibration issue."

Model thin and with the unglamorous look of a doctor who spends her time caring for the tiniest people on earth rather than applying mascara, Dr. Van Marter kept a poker face. "What do you mean by an equilibration issue?"

The resident was unprepared to get called on this offhand comment. "Um, well, I was under the impression that anemia is often seen after delivery, right?" She looked at the fellow for support, but the fellow was scribbling notes on her rounding sheet and leaving the teaching to Van Marter.

"Yes, it is common to see a low blood count in the NICU," Dr. Van Marter answered the resident. "But when do we typically see anemia? It's not on the second day of life, is it?"

"Around a week?" The resident looked like she was guessing.

"Yes. So you would recommend . . ."

"Repeating the blood count?"

"This afternoon."

"Absolutely."

Van Marter was puzzled. She had watched Larissa's delivery and had been the one to perform the intubation. The colleague who had covered the NICU overnight had reported nothing out of the ordinary when she'd signed out the patients to Van Marter that morning. Larissa's first few hours had been relatively smooth—at least as smooth as could be expected for a baby born at twenty-six weeks.

But Larissa's blood count was too low. Van Marter made a note to watch out for the results of the repeat blood test after lunch.

"So," Van Marter said, turning to the medical student in the group and giving a wink to the resident she had just finished questioning. "What are the possible causes of a low blood count?"

"Well," said the medical student, stalling. "ABO incompatibility could cause a low blood count." The student looked like he was trying to conjure up a page of his textbook in his head. The pockets of the short white lab coat that needed a cleaning were stuffed with papers and reference books.

"Yes, it could. How common is ABO incompatibility?"

The medical student shook his head.

"Zero point six percent," said Van Marter, indicating the rarity of this condition in which the mother's immune system attacks the baby's red blood cells. "And do you know what percentage of babies have a low blood count as a consequence?" The student again shook his head. "A very, *very* small percentage." Van Marter enunciated each word. "Any other thoughts?"

"It could be caused by bleeding."

"Excellent. What type of bleeding?"

"Blood loss at delivery?"

"Yes; where else?"

"Internal bleeding."

"Uh-huh, and where? Where do premature babies bleed?"

"Internally?" The medical student apparently hadn't read this section of his text, and he stuffed a bony hand into the pocket of his slacks.

"Okay . . . How about studying the causes of neonatal anemia and reporting back to the team tomorrow at morning rounds? I guess we know what you need to read about tonight."

The group moved to the next baby.

With less difficulty than she'd had the previous evening, Kelly got into her wheelchair and we rolled to the NICU. We now knew that when we passed into the alcove between the locked outside door to the NICU and the door that opened into the vast and busy space, we were expected to stop and wash our hands, which we did.

Maneuvering the wheelchair to Larissa's station, I couldn't help noticing that the space that had been blocked by the blue privacy partition the previous evening was now empty.

Overnight, Larissa had been moved off the open table and into a Plexiglas-enclosed incubator. She lay propped on her side, snuggled between rolled-up blankets and moored by the semirigid ventilator tubing.

I rolled my wife's wheelchair up to the incubator, and Kelly readily unlatched the two portholes and reached in to touch our daughter. We pressed our faces to the plastic and watched Larissa. Kelly gently rubbed Larissa's hand and smiled when Larissa reacted to the touch.

"She had a good night." The nurse smiled at us when she said this.

"Did you have a good night?" I asked Larissa.

"Oh, that's good." Kelly did not seem entirely ready to believe it.

Later, as we were leaving the NICU, Dr. Van Marter peeled off from the group she was huddled with, caught up with us, and introduced herself. She offered her congratulations on the birth of our daughter and told us she was optimistic because Larissa's first day had gone well.

"Thanks so much for everything," Kelly said. She was looking tired and a bit pale. I started to turn the wheelchair toward the door.

"You know"—Van Marter interrupted my motion with her voice—"her blood count was a little low this morning, and I repeated it and it's still low, so I'm going to get an ultrasound just to be sure."

"An ultrasound of what?" Kelly asked.

"Of her head," Van Marter said without sounding too concerned. "It's probably nothing, and usually we don't get an ultrasound of the head until babies are three days old, but why not? Let's go ahead and get it now."

"Sounds good," I said with a smile, but suddenly I felt beads of perspiration and heard a roaring sound in my ears.

"I'm here until tomorrow," Van Marter said, "so we can touch base later." She walked off. I pushed Kelly's wheelchair out of the NICU.

"What's the ultrasound for?" Kelly asked. "What are they looking for?"

"Bleeding. An intraventricular hemorrhage. They call it an IVH."

"Do you think she has that?"

"I doubt it," I lied. I could see no reason to worry Kelly with another issue that was out of our control. "I'm sure Van Marter is just being careful."

Kelly wanted to take a nap, so after getting her settled in bed, I headed for the library. Four stories above the Brigham and Women's lobby, tucked behind an inpatient ward, the library served a lot of purposes—a meeting space for residents, a place to check e-mail and, when time allowed, catch an hour of sleep on one of the comfortable, if slightly dingy, couches. I swiped my hospital ID to open the door and then sat at a computer terminal between a young doctor writing a scientific paper, his notes strewn haphazardly around him, and a medical student intently engaged in an instant-messaging exchange. Around me, tired residents scanned patients' labs, checked e-mail, or looked for cheap flights to vacation spots.

I was there to do research on intraventricular hemorrhages. I searched for studies that showed how premature babies who'd had severe hemorrhages near birth did in childhood. There were only two studies, and neither was encouraging: a small number of children suffered only mild disabilities, but most were profoundly affected—with mental retardation, seizures, cerebral palsy, and other disorders.

The brain is composed of billions of cells that make trillions of connections with one another. At the highest level is the cortex of the brain, which communicates with the body by sending signals down the spinal cord.

Think back to your high school science classes and remember the picture of the brain: two hemispheres encasing a pair of fluid-filled cavities, called ventricles. The fluid in the cavities, which is indistinguishable from water to the naked eye, is cerebrospinal fluid. It bathes and cushions the brain by constantly circulating through a sequence of narrow channels that connect the ventricles to the space surrounding the brain and the spinal cord.

Along the ceiling of each ventricle lies a fragile network of blood vessels called the germinal matrix. This immature matrix tends to bleed after a premature birth, spilling blood into the fluid-filled ventricles. Once blood accumulates, in the ventricle or anywhere else, it clots.

Scanty bleeding is clinically meaningless, as the drops of spilled blood are reabsorbed over time. In fact, minor bleeds are so common that if every full-term baby had an ultrasound of the head, up to 10 percent of them would be found to have evidence of bleeding—a testament to the irrelevance of a truly small hemorrhage.

Significant bleeding into the ventricles is an entirely different matter. A large clot can clog one of the narrow channels and obstruct the flow of fluid, causing pressure in the ventricle to build. When pressure inside a ventricle increases, the ventricle expands, strangulating the blood supply to the brain cells that line the ventricle, slowly killing them. Injury—to brain cells, to the spinal cord, even to the cells that insulate the neurons—interrupts the connections that are central to the brain's function, and broken connections result in impairment of movement, thought, and behavior.

Significant bleeding can also occur within the brain tissue itself, and an increasing amount of blood compresses the surrounding brain tissue, killing nearby brain cells and leading to permanent disability.

Hemorrhage in the brain is diagnosed with an ultrasound. The dishwasher-size machine, just like the one used to view the fetus during pregnancy, is wheeled to the bedside, and the technician snakes a handheld probe into the incubator. The probe is positioned against the newborn's head; ultrasonic waves are sent inside the skull, and the sound waves that bounce back are analyzed by computer and then transformed into an image that's displayed on the screen.

Later that night, after Karen, Hannah, and Grace had visited and while Kelly was watching an *Ally McBeal* rerun on television, I went to the NICU to inquire about the ultrasound.

I found Dr. Van Marter discussing some lab values with one of the fellows. It was late, and the NICU was quiet; the lights were dimmed, and a smaller staff tended to the babies overnight. The neonatologist saw me and motioned for me to wait while she finished her discussion, and then she came over to me. "Let's go someplace quiet," she said, touching my arm to lead me out of the NICU.

She took me down a hallway to a conference room, where we sat alone at the end of a long table.

"I know Larissa lost some blood in the operating room," she began in a balanced tone. "So I wasn't too surprised when her hematocrit was low. But I expected it to come up when we transfused her."

"I didn't know she was transfused—" I began.

"She was." Van Marter silenced me, then returned to her topic. "The blood count came up, but not as much as it should have. You probably know that the brain is the most common place that premature babies bleed, which is why I ordered the ultrasound study." When a significant intraventricular hemorrhage occurs, red blood cells pour out of tiny vessels and accumulate in the ventricular fluid; lab tests then show a lower number of red blood cells in the circulatory system.

Van Marter looked down, and then looked up at me. "The ultrasound showed a bleed. A very serious bleed. Though not the worst I've ever seen, Larissa's hemorrhage is on the more severe end of the spectrum."

On an ultrasound image, fluid in the ventricle appears as a dark half-moon within the uniform gray of the surrounding brain tissue.

Blood appears as a shocking white cloud within the black of the ventricle or against the gray background of the brain tissue.

The ultrasound confirmed Dr. Van Marter's diagnosis and one of my worst nightmares. The initial scan showed that blood had flooded the ventricles, threatening to raise the brain's internal pressure and compress the healthy brain tissue. It also showed blood in the tissue surrounding the ventricle on the left side of Larissa's brain, as well as in the cerebellum, the small sea-urchin-shaped portion of brain that juts out from underneath the two hemispheres at the back of the head. The cerebellum is the coordinator of the brain; it organizes and synchronizes much of the brain's thoughts and functions, and hemorrhage here is usually a catastrophic event.

Last, the ultrasound showed that the brain had begun to swell—fluid had built up in the left ventricle, and the ventricle was expanding into the space that ought to be occupied by the left hemisphere of the brain. There was even evidence that the membrane that marks the midline of the brain was shifting toward the right, meaning the whole left brain was swelling out of its natural space and encroaching on the right side of the brain in an effort to accommodate the increasing size of the left ventricle.

The radiologist who'd looked at the images and dictated the report described what essentially amounted to an explosion of blood in Larissa's head. We were seeing the beginning of a cascade of events inside the skull whose outcome was unknown.

A day after her birth it was clear that the term *unscathed* had been taken off the table. After my library session, I knew the consequences of Larissa's ultrasound findings: she would likely have catastrophic movement impairment similar to severe cerebral palsy and affecting all four limbs. It was unlikely that she'd have normal intellectual function. Her early life would be marked by neurosurgery to relieve increasing pressure in her brain and orthopedic surgery to relieve the consequences of muscles unable to relax. Body functions and developmental milestones most of us take for granted—swallowing, seeing, hearing, sitting up, learning to speak—might be achieved slowly, or not at all.

Because of the newborn brain's plasticity—meaning its remarkable ability to reassign tasks from an injured area of the brain to a healthy one—we had no idea where on the continuum Larissa would fall. An adult with Larissa's ultrasound findings would certainly die or be severely and permanently disabled, and either outcome was a real possibility for Larissa. However, unlike adults, newborns have robust neuroplasticity—brain plasticity—so Larissa's uninjured brain cells might eventually compensate for the brain tissue that the ultrasound showed was now a bloody morass.

Meanwhile, her immediate survival was in no way guaranteed, particularly given the evidence that bleeding on the left side of her head was causing compression of her brain. Dr. Van Marter explained that Larissa would have daily ultrasounds to track the progress of her hemorrhage.

"We will have to see whether the hemorrhage stabilizes or gets worse over the next few days," she said. "If she bleeds more, there may be some hard choices to make on Monday or Tuesday." Then she added, "But if the bleeding continues, Larissa may make those choices herself."

I followed the pediatrician's train of thought. "If there is a decision to be made, is it mine and Kelly's to make?"

"It will depend," she answered. "Let's take it one day at a time and see what Larissa tells us."

In a daze I walked back to Kelly's room. I passed a colleague who asked me how Larissa was doing; I walked by without acknowledging his presence. I felt numb—not sad, or anyway not despondent, but unable to process, unable to consider what was happening. It was all I could do to make it back to Kelly's room.

"What did she say?" Kelly asked. She had been waiting for my return. I wasn't sure how long I'd been gone.

I shook my head. Kelly started to cry as I sat down on the bed next to her.

CHAPTER 3

Gifted Hands

The holiday season was a busy time for the chief of the largest NICU in New England. Dr. Steve Ringer was tall, not lean, and wore a thick beard, making this Jewish physician from the Boston suburbs ideal for the role he played every December: Santa. One December morning found him handing out gifts at the Department of Neurology's holiday party as children climbed on his lap and whispered in his ear about toys.

Five floors above, a woman who had been hospitalized for three weeks due to concern that she would deliver her baby prematurely called her nurse and explained that she felt like she needed to pass a bowel movement. Because that sensation is not dissimilar from what a woman may feel right before her baby is born, the nurse panicked, told the patient she needed to wait until a doctor examined her, and raced back to the nursing station to call for help.

The young resident physician who arrived didn't recognize this sign of trouble, and she allowed the patient to proceed to the toilet, where she unceremoniously delivered a vigorous three-and-a-half-pound baby boy into the bowl.

A few minutes later, Santa's pager went off. Ringer raced upstairs in full red regalia and arrived just in time to see the little boy fished out of the toilet. As the shocked mother dropped back onto the toilet seat, Ringer helped dry the baby and told her he was pleased to report that the newborn was doing well. Soon the neonatology team caught up with their leader, bundled the new arrival into an incubator, and whisked him off to be evaluated. After congratulating the mother and assuring her that her baby looked very healthy, Ringer returned to the party.

Ringer was, in some ways, an unlikely leader of the NICU. He had grown up in the Boston suburb of Newton, not far from where he currently lived. He had watched his father, an internal medicine doctor, spend long days seeing one patient after another. It seemed to Ringer that although physicians occupied a place of distinction and respect in the community, their actual work was somewhat boring. In a community where Harvard Medical School cast a long shadow, Ringer thought that science—with its potential for lifesaving discoveries—was a more interesting destination than the exam room and so he had immersed himself in laboratory work from an early age.

Ringer went to college at Brandeis (fifteen minutes from Newton), where the laboratory opened up to him. He got a taste for experimenting in biology and chemistry and came up with a plan: he would become a physician scientist.

With his wife, Ellie, Ringer moved to Cleveland, where he went to medical school and got a doctorate in biochemistry. "I spent a lot of time shining lasers through solutions to measure the mass of large molecules of DNA," he recalled. He decided on pediatrics and started his residency at Case Western Reserve.

Residency was a time of false starts. His experience in the clinic treating runny noses and earaches confirmed his suspicion about general practice: it was boring. In order to avoid the boredom, he needed to subspecialize. The question was, What subspecialty should he choose?

Because of his interest in research, he thought that clinical genetics would be his future, but a month-long rotation piecing together complex diagnoses from the constellation of unusual clinical findings found him bored yet again. A friend offered some advice: "When you find yourself browsing shelves in the medical library, pay attention to the journals and articles that hold your attention."

"I started paying attention," Ringer said, "and I came back to him and said, 'I'm reading these papers in genetics, like a baby with a novel form of something or other or a baby with a particular enzyme deficiency. I don't get [why I'm bored with clinical genetics]. I'm reading all of these genetics papers.' "

"No, you idiot," the friend responded. "You're reading articles about babies."

Never a slouch during college and medical school, but not at the top of his class either, Ringer found residency to be more suited to his temperament. In an environment marked by grinding days and nights, sick children, anxious parents, and unpredictable colleagues and supervisors, Ringer stood out. He had gifted hands and a way with people; in a high-stress atmosphere, these were critical skills. By the time he chose neonatology, his reputation had grown, and every training program wanted him.

The previous year, Ringer had done a summer research elective in a lab at the Marine Biological Laboratory on Cape Cod, working with the sperm and eggs from sea anemones. It was a plum assignment and in a place where his young family could spend a month at the beach in lab-subsidized housing. The lab was a side interest of Bill Speck, chief executive officer of the pediatric hospital in Cleveland where Ringer was a resident, and Speck would visit once or twice each summer to check on his research.

The following summer, as Ringer was deciding which fellowship to take, he returned to Speck's lab on Cape Cod. Unseasonably warm oceans had interrupted the sea anemones' gamete production, and when Speck visited, Ringer had to tell him that there was nothing to research at the moment. Speck had just bought a boat, and so while they waited for the anemones to spawn, the two Cleveland doctors—one an amiable resident, the other the CEO of the hospital—took the boat out and tried to figure out how to catch a striper, which, based on conversations they'd overheard at the dock, was the fish to catch. (Once they caught a bluefish—a dark-fleshed, strong-tasting fish not particularly appreciated outside of New England—and assumed that the dark lines along its glistening skin must be the stripes they'd been in search of. It took the smirks at the dock to disabuse them of the pride that they had caught a striped bass.)

While they sat in the boat under the bright summer sun, they got to talking, and Ringer heard amazing tales of infighting and political maneuvering at the highest levels of the hospital. To the young and impressionable Ringer, hospital administration seemed fascinating and even glamorous.

Ringer moved back to Boston and began a fellowship at Harvard's Joint Program in Neonatology, an intellectually rich program that linked the academic resources of Harvard Medical School with clinical training in several of the most prestigious Boston hospitals, including Brigham and Women's. In 1988, when Ringer and Ellie moved into a small home in Needham, the next town over from his boyhood home, neonatal medicine was enjoying the start of an arc of innovation that bent steeply up.

Sporadic attempts to sustain the lives of babies born too early dot the history of medicine. In the 1880s, the first incubator, designed to help premature newborns maintain their body temperature and modeled on warmers for poultry eggs, was developed, and a few years later, a French physician reported success with a flexible tube inserted in the nose that allowed milk to be delivered directly to the baby's stomach.[1]

The oddest chapter in the history of neonatology started in the 1890s, when premature newborns, typically born six to ten weeks early and weighing between two and three pounds, began to appear at expositions across Europe inside the newly invented incubators. People flocked to see these marvels of science sustained in strange, glass-enclosed contraptions. By the early 1900s, these incubator-baby exhibits were regularly featured at state fairs and science demonstrations in the United States; a permanent display at Coney Island in New York lasted until the late 1930s. Rigorous statistics were not kept, but medical lore has it that the incubators increased survival for these babies to about 75 percent (most of them would have died otherwise). Once the babies reached a normal birth weight, they went home.[2]

Aside from that innovation, and subsequent spectacle, prematurity was one of those unavoidable and largely untreatable problems. If a baby was too small to breathe and couldn't digest milk or formula, there was little the pediatricians could do. Perhaps because of the futility of treating it, prematurity was not a high-profile problem; this meant that for decades, any young academic physician seeking to make a name for himself chose to focus on curing cancer or pioneering lifesaving surgeries, not saving babies born weeks too early. Special units for premature newborns were established at a few hospitals, including the Children's Hospital of Philadelphia—future surgeon general of the United States Dr. C. Everett Koop established a NICU there in 1962—but at most hospitals, premature newborns were placed in the pediatric intensive care units alongside youngsters with asthma, burns, head injuries, and other life-threatening injuries and illnesses.

The birth and death of Patrick Bouvier Kennedy changed that. The youngest son of President Kennedy, Patrick was born by cesarean section on August 7, 1963, at thirty-four and a half weeks, weighing nearly five pounds. Although today this degree of prematurity would be trivial, Patrick developed respiratory distress syndrome soon after his birth and was transferred from the air force base in Bourne, Massachusetts, where he was born, to Children's Hospital Boston.

Overnight, the medical system's limitations in caring for premature newborns were starkly revealed.

Robert Levine, considered to be an expert in premature infants at the time, was walking his dog outside his apartment on Central Park West in New York City when a police car pulled up beside him and an officer told him to get in. Levine asked why and was informed that the president needed him. He was taken by helicopter to Boston, but he was able to offer practically nothing other than consolation.

Maria Delivoria-Papadopoulos, a Greek pediatrician training in neonatology in Toronto, had modified an adult ventilator for use in preterm newborns, and months earlier she had published a paper describing how she had saved the life of a thirty-four-week baby girl who had developed lung disease shortly after birth. Desperate, the doctors at Children's Hospital contacted Delivoria-Papadopoulos's supervisor at the Hospital for Sick Children in Toronto and discussed flying Patrick Kennedy to Canada for treatment, but ultimately this was ruled out due to political considerations.

Without any other options, the staff at Children's Hospital

Boston put Patrick in a hyperbaric oxygen chamber—the same apparatus used for burn patients and divers with the bends—but to no avail. He died on August 9.

Levine, who had come to Boston in his pajamas and an overcoat, had to borrow money from a friend so he could buy a train ticket back to New York.

More than any new technology had, this death motivated pediatricians. Over the next few years, intensive care units designed specifically for newborns—full-term and preterm—sprung up at hospitals with large delivery volumes.

Now discoveries in newborn medicine came rapidly, as researchers examined the way each organ failed to function prior to the completion of development. Feeding presented two challenges: swallowing and digesting. Swallowing is a complex mechanical operation requiring the coordinating of mouth, tongue, and throat musculature to move liquid from mouth to esophagus without allowing it to enter the lungs. Although there are exceptions, many babies can't master bottle- or breast-feeding until they are around thirty-four weeks of gestational age.

A simple way around this was tube feeding, pioneered as early as the 1890s, in which a tube was passed through the baby's nose and down the esophagus, avoiding the problem of swallowing altogether and allowing nutrition to be dripped into the stomach.

This approach was lifesaving for newborns that could not yet swallow, but it still relied on the intestines to function normally. Most basically, the job of the intestines is to digest food into soluble proteins, carbohydrates, and fats and move them into the bloodstream, where the nutrients are transported throughout the body. Severely premature newborns could often be coaxed to digest milk or formula, but the process was slow and painstaking, and at a time when babies needed to gain weight fast, their intestines were notoriously unpredictable. Pediatricians had to bypass the intestines; they needed a predigested form of nutrition that could be infused directly into blood vessels. This seemed straightforward enough, but a host of obstacles complicated the development of this relatively intuitive technology.

Researchers created cocktails of carbohydrates and proteins and infused them into research subjects' veins, only to find that the solutions were so concentrated they caused chemical injuries that obliterated the veins.

Next, investigators diluted their solutions with large volumes of saline in order to protect the veins and still get the nutrition into the patient. Reports in the 1940s and 1950s described adult patients infused with five to seven liters of fluid, causing nearly constant urination.

The breakthrough came in the 1950s with the discovery that if one advanced a catheter into a major vein at the center of the body—the vena cava, near the right atrium of the heart, became a favored location—the large volume of blood rushing through diluted the nutritional infusion and prevented injury to the blood vessels. This discovery allowed doctors to infuse concentrated solutions that provided sufficient calories for survival; the massive volumes of fluid that had been used in earlier years were no longer needed.

The first babies to receive intravenous feeding solutions in this way—called total parenteral nutrition, or TPN—were those born with gastroschisis, a relatively common condition in which the abdominal wall fails to form completely and the baby is born with its intestines exposed. These babies, who were in every other way normal, had been making it through the surgery required to close their abdomens but then dying of malnutrition before their intestines woke up and started functioning normally after their intestines had been replaced surgically inside their abdomen. (In babies, as in adults, the intestines often temporarily stop working when irritated by surgery or any other noxious stimulus.)

The use of TPN in newborns with gastroschisis turned a condition that had been almost uniformly lethal into one that was nearly always survivable. Over the course of a few short years in the 1960s, the death rate for babies with gastroschisis dropped from more than 75 percent to less than 5 percent.

It didn't take long for the pediatricians to realize that TPN could also be used to help extremely premature newborns gain weight and grow. And as the TPN solutions became more sophisticated and more "lifelike," the complications of using a substitute nutrition—liver failure and infection were the two most common—became less frequent.

Other discoveries were less dramatic. Humidified incubators

helped regulate premature babies' body temperature and limited fluid losses through their semitranslucent skin. (When one unit director in South Africa couldn't afford incubators for his NICU, he turned the thermostat up to body temperature—98.6 degrees—and allowed the staff to come to work in shorts and T-shirts.) During these years, miniaturization of adult technology—from ventilators to catheters—provided pediatricians with equipment for the specific needs of their tiny patients.

Another major discovery, this one outside of the NICU, was that giving women in preterm labor steroid injections substantially reduced lung disease and brain injury in their premature infants. These injections became widespread beginning in the 1980s and 1990s, and those babies who were lucky enough to be exposed to the steroids while still in utero did much better after they were born than those who weren't.

Perhaps the most important innovation in the care of premature newborns began with a 1959 discovery by a young pediatrician with an interest in prematurity. She'd been told to go learn about the foam that frothed from the mouths of adults with pulmonary edema, a lung disease usually resulting from heart failure, and so by day Mary Ellen "Mel" Avery, who would later become the top doctor at Children's Hospital Boston, studied adults who foamed at the mouth. By night, to supplement her meager research stipend, Avery took care of babies born at the Boston Lying-In Hospital. Curiosity led Avery to search for a link between adults with pulmonary edema and preterm babies with lung disease, and the Lying-In gave her access to autopsy specimens of babies who had died of lung disease. What she found was startling: in those premature newborns, the tiny air sacs that were surrounded by blood vessels at the periphery of the lung had collapsed. Because the air sacs in the lungs didn't stay open, the critical exchange of oxygen and carbon dioxide didn't take place, and the babies essentially suffocated. Avery discovered that the babies' lungs were collapsing because the surface tension holding the damp air sacs closed was insurmountable.

To understand what surface tension is, try this: First, lift a dry paper towel off a tile floor; you'll notice that it comes up easily. Next, soak the towel in water, spread it on the floor, and try to pick it up. The "stickiness" between the damp towel and the floor is surface tension, and the same mechanism causes the damp walls of tiny air sacs to stick together. In a 1959 paper, Avery demonstrated that the surface tension in the lungs of premature babies who had died of lung disease was more than three times higher than the surface tension in the lungs of babies who had died of other causes. She speculated that babies with lung disease lacked some substance that reduced the lungs' surface tension and prevented them from collapsing. The search was on for surfactant—the word comes from a contraction of "surface-active substance"—which is a natural product of the lung tissue in full-term babies.

At first no one believed Avery. "Mel's playing with soap bubbles again" was something she heard repeatedly. But in the 1970s, her research began to come together: the composition of surfactant was worked out, the lung cells that made it were identified, and the specific effect the substance had on the lung tissue was demonstrated. (It turned out that the foam coming from the mouths of adults with pulmonary edema was full of surfactant.)

Avery teamed up with Japanese researcher Tetsuro Fujiwara, and, in 1980, an article announcing their research findings was published in the *Lancet.* A picture conveyed whatever their words did not. The researchers instilled surfactant into one lung and took an X-ray, and that X-ray is reproduced in the journal article: one lung is inflated, and the other is collapsed.[3]

It took until 1991 to get animal surfactant approved for use in premature newborns.

"It was an incredibly exciting time," remembered Michael Epstein, who preceded Ringer as the director of the Brigham and Women's NICU. "The innovations and discoveries came very fast."

There were errors as well. Focused on lung disease and the problem of getting oxygen to these newborns, doctors had taken advantage of new incubators that allowed for the delivery of much higher oxygen concentrations to the babies. They didn't spend a lot of time wondering why babies with lung disease were more likely to go blind from retinopathy of prematurity than babies without lung disease. It wasn't until the 1990s that they realized that retinopathy of prematurity was caused by too much oxygen, and that premature babies thrived optimally on oxygen levels much lower than the levels required by full-term babies and adults. When they realized this, doctors started dialing down the oxygen concentrations, and they were dismayed to see the incidence of cerebral palsy rise as the reduced oxygen levels caused brain damage. Nuance was required, and babies suffered the consequences of the new field's evolving science.

But during this generally golden era of newborn medicine, from 1963 until the mid-1990s, the death rate for premature newborns dropped fast. One study drawing on data from 1958 to 1968, the early era of neonatology, reported that the mortality rate for babies born at twenty-eight weeks was 70 percent.[4] By 1988, when Ringer arrived in Boston, the mortality rate for babies born at this gestational age had fallen to 10 percent.

From his vantage point in Cleveland, Ringer thought the Harvard Joint Program in Neonatology looked perfect. He could gain clinical skills at the three great NICUs that made up the program, and he would have an entrée into the robust research community at Harvard. Ringer felt confident that he could launch his research career in that environment.

What Ringer didn't realize was that newborn medicine at the Harvard hospitals was chaotic. Physicians rotated from hospital to hospital, each of them bringing his own "very best" way to care for sick newborns. And because the field was so new, there wasn't a body of medical evidence to point physicians in the right direction. Ventilator settings were arbitrary, infections were common, surgeons came and whisked babies off to the operating room without explaining why.

The nurses—who did not rotate among hospitals—were left to sort out the disorder and try to maintain consistency. Animosity developed: nurses did their best to defend the babies from the decisions of the transient physicians, and the physicians did their best to bring what they believed was the most optimal care to the babies.

At Brigham and Women's, hospital administrators had the brilliant idea of putting the NICU in the basement. Physicians who worked there remember having to lean over the secretaries' desks in the reception area and crane their necks upward toward the single skylight to see if it was snowing or raining, or whether darkness had fallen. The babies may not have minded the lack of windows, but the location added to the dysfunctional working environment.

Ignorant of the chaos as a trainee, Ringer found a lab that was studying the function of white blood cells, and he settled in to grow his career as an academic researcher and neonatologist.

Amid the babies, Ringer knew he had made the right choice. In his oversize beefy fingers, intravenous catheters found their way into the tiniest of veins, and breathing tubes reached just the right spots in the tracheas of the smallest premature newborns. He had an intuitive sense for the babies and quickly developed a feel for which of them would thrive and which would not. The nurses, a generally skeptical lot, trusted Ringer. They made him cookies, they laughed at his jokes (which were sometimes even funny), and they came to him to arbitrate conflicts.

"We should talk," he would say to the parents of a child who was deteriorating, and he had a remarkable ability to connect with those parents and make them understand. Nurses loved that he could do that.

But his research did not flourish. He struggled to put in the time in the lab, since he was drawn more to the bedsides of his babies than to his tubes and Western blots. Research papers, one of the key currencies of academic success, took forever to write. And his research mentor there, who had never been particularly attentive or predictable, went screaming out of the laboratory one day and never came back. Rumors circulated about a psychotic break.

Feeling somewhat dispirited but extremely well trained, Ringer began to make plans to return to Cleveland. His clinical acumen and his exposure to cutting-edge research at Harvard made him attractive to the NICU in Cleveland where he had done his residency, and his research mentor from his years in Cleveland invited him back to the lab.

Just months away from completing his training, Ringer wandered in to Michael Epstein's office to let him know his plans. Epstein was the head of the Brigham NICU and was about to become the director of the Harvard Joint Program in Neonatology—a hard-charging doctor who, like Ringer, had a knack for threading intravenous lines into the tiniest of veins.

Epstein too had aspired to a research career because that was what people said was the highest calling in medicine, but he'd found he didn't like it very much and wasn't particularly good at it. He had ridden his clinical skills and his capacity to manage people to a position of leadership in the nascent newborn-medicine community.

Epstein had watched Ringer during his years in Boston and thought he saw something familiar. "I looked at Steve and saw myself," Epstein recalled. "He was terrific clinically, and he was beloved by the residents, the medical students, and the nurses. He could teach at the bedside and take care of sick babies. And he had a sense of humor, which I believe is about ninety percent of success in life."

Ringer told Epstein he planned to return to Cleveland.

"That's great. Congratulations," Epstein told him.

"Well, thanks," Ringer said.

"You don't seem very happy about this," Epstein observed.

"Well, actually, I'm not very happy. My research hasn't been that successful. The people in Cleveland are supportive, but I'm not that confident that my research career is going to work out."

"Well," said Epstein, "would you ever consider running a clinical unit?"

Today, it would be unthinkable to take a young doctor straight out of a training program and put him in charge of the largest newborn intensive care unit in New England, but in 1988, the field was young, and hospital administrative structures were less formal. Ringer went from trainee to director overnight.

Ringer remembered what he'd thought it would be like to run the NICU. "My impression was that Michael [Epstein] would come out of his office at one o'clock every day and walk around the unit," Ringer said. "He would talk to the nurses, ask how things were going, and be gone by one ten or one fifteen. I thought to myself, *What a deal. I'm going to work twenty minutes each day and be twice as good as him. I'll sit in my office the rest of the time and read papers and get very smart.* Of course it didn't turn out that way."

The truth was that the NICU was in chaos. The nurses knew it, and the hospital's administration knew it. And it became Ringer's job to fix it.

Sunday dawned gray and cold. I slowly awoke in the half-light, looked up from the cot where I'd slept to the hospital bed where Kelly dozed, and the awfulness rushed back. I felt the tightness return to my stomach.

We made our way to the NICU and to the incubator where Larissa lay, just forty-eight hours old. There was less bustle around her now. The incubator had acquired a name tag, decorated with cute curly letters in a few colors, and a couple of photos of Larissa were taped to the wall near the bed. I pushed Kelly's wheelchair closer to the incubator, and she opened the two portholes and stretched her hands inside to stroke Larissa's silky thin hair.

The air in the incubator was warm, and Larissa was dressed in a tiny white onesie. She was positioned on her side, supported at her back by a tiny wedge-shaped pillow. A warm cotton blanket covered her. The scene looked almost normal except for the rigid plastic tube that emerged from Larissa's mouth and was taped in place at the sides of her head. The tube was connected to coiled plastic tubing that exited the incubator and arced down to the ventilator, which was emitting its *whoosh-whoosh* sound.

"Mommy loves you," Kelly said.

Dr. Elisa came over, her manner almost relaxed.

"She had a good night."

"That's good." Kelly smiled.

"Her vent settings are pretty low. We might even be able to get her off that soon."

We stared at her.

"And her blood count seems to have stabilized."

"That's good?" Kelly ventured, looking at me.

"She's on antibiotics as a precaution, but she isn't behaving like she is infected."

The three of us stared at Larissa through the Plexiglas as Kelly stroked her head.

We gazed down at a gorgeous tiny child lying peacefully between clean cotton pillows and blankets. Save for the stark green translucent tube jutting from her mouth, Larissa looked fine.

The problem was that we knew she wasn't fine, that inside her head was a blood clot pushing aside fragile brain structures that now would never develop normally. It was like a clean shirt placed on a gunshot victim—the apparent tranquility was false.

Elisa had retreated, and after a minute I followed her, leaving Kelly at Larissa's side.

"So," I asked Elisa, "what decisions have to be made?"

"I'm not sure what you mean." Elisa looked genuinely puzzled.

"For Larissa."

"I'm sorry." Elisa was trying to understand me. "I'm not following you."

"Dr. Van Marter spoke to me about decisions that needed to be made if Larissa deteriorated."

"To be honest with you, I'm not sure there is much to decide," Elisa said. "Larissa is really pretty stable. Sure, she's on the ventilator, but I expect her to come off that pretty soon, and other than that, she's getting the usual support: antibiotics, fluid, nutrition. I think Dr. Van Marter was talking about decisions that would have to be made if she got worse. To be honest with you, I think she looks pretty good," Elisa finished. "Don't you?"

"But what about her bleed?" I asked. "She has a grade 4 IVH and a cerebellar hemorrhage. What will that mean? What kind of a life will she have?"

"I'm not sure," Elisa said. She smiled kindly but her tone had taken on a hint of steeliness. "We will ask a neurologist to come evaluate Larissa on Monday.

"But for now I'm pretty sure there isn't much to be done." She wrapped up the conversation.

I rejoined Kelly, who was now in pain, and we returned to her room so she could lie down and take a Percocet.

After lunch, we heard the sounds of children in the hallway, and a moment later, Grace and Hannah burst into the room, Karen close behind.

"Mommy! Mommy! Mommy!"

Hannah in particular had to be restrained from climbing into the bed with Kelly, who steeled herself against the pain of her cesarean incision and smiled at her girls.

"Where's Larissa?" Hannah asked, not understanding why her sister wasn't in the room.

They checked out the motorized bed, the hospital-issue soap, and Kelly's gown. Then they wanted to see Larissa, so Kelly climbed back into the wheelchair, and we took the elevator to the sixth floor.

"What gorgeous girls," the secretary said to Hannah and Grace while we were waiting for the okay to go into the NICU. "And what good sisters you are," she said, pointing to the cards and drawings they had made for Larissa.

Larissa's nurse appeared, and we were cleared to enter the NICU. Kelly and I kept exchanging glances. What would the girls think? Would they be scared? Would they want to leave? What would we say if they asked about Larissa's future?

"Do you girls know how to wash your hands really well?" the nurse asked Hannah and Grace. They looked at her patiently and scrubbed their hands in the big sink.

On our way through the NICU, we kept looking at the girls for signs that they were overwhelmed or scared by the beeping machines and the activity. But if they were anxious, their desire to meet their sister kept it in check, and they walked in single file after the nurse to Larissa's bedside.

"Oh, look, she's so cute," Grace said.

"I can't see," Hannah complained. "Lift me up.

"She is cute," Hannah agreed after I picked her up so she could see inside the incubator.

"Where did she get that cute outfit?" Grace wanted to know.

"She got it here in the NICU," the nurse explained.

"Can we get her some clothes?" Hannah asked.

"Can we touch her?" Grace asked. Carefully, one of the portholes in the incubator was opened, and the girls took turns touching Larissa's foot.

"She's very soft," Hannah noted. "Larissa, want to see the pictures we made for you?" Hannah asked when she was done.

The girls taped their pictures up near Larissa's incubator, looked at the photos, and admired Larissa's name tag. Then they wanted a snack, so we headed out.

"What's the tube for?" Hannah wanted to know, almost as if it were an afterthought, as we started to leave the unit.

"That helps her breathe, right, Mommy?" Grace said.

A group of nurses stopped the girls.

"Did you visit your brother or your sister today?" one of them asked Grace.

"My sister Larissa," Grace explained as we came up behind them. I pushed Kelly in her wheelchair.

"She's gorgeous, isn't she," another nurse said. "Just like the two of you." She looked up at me.

"You have a beautiful family," the nurse said.

We were headed for the Au Bon Pain in the lobby to have chocolate croissants and hot chocolate. A woman pushing a washing-machine-size ultrasound rolled past us, moving in the general direction of Larissa's bed.

"Baby Lowery?" I heard her ask.

One of the staff pointed absentmindedly in Larissa's direction.

"The dad works here?" one of the nurses asked.

"I heard he's an OB intern," another said.

"Lord help us," a third said. "A little knowledge is a dangerous thing."

"Hey, Elisa," one of them said, roping Dr. Elisa, who was walking past, into the conversation. "Is it true the dad wants to take her home?"

"Well, I don't know about that," Elisa said.

"I heard he was asking you about his rights."

"I think he's just concerned," Elisa said. "Who wouldn't be concerned?"

"He's got a picture-perfect family."

"Now, now," Elisa scolded, and continued making rounds.

CHAPTER 4

"The Degree of Impairment Is Difficult to Predict"

On Monday, January 14, Larissa's fourth day of life, Ringer came to work feeling rejuvenated. The holidays were behind him, and the weekend had been calm. It was bright and cold as Ringer walked up the shallow hill between the employee parking lot and the hospital. Once there, he took the elevator to the sixth floor.

"Good morning," he called to the receptionists who served as gatekeepers to the NICU. He rounded the corner and entered the unit's administrative suite, greeted his secretary, and went into his office. By any standard, the office was a disaster. Papers and textbooks covered his desk, creating a layer at least two inches thick. A small table in the office had a pile of papers almost a foot high. Stacks of books filled two chairs, covered a file cabinet, and formed columns right outside the door, the result of a fruitless triage effort some months back. Ringer dumped his coat on top of the pile on the file cabinet and logged in on his computer to check his e-mail.

"I need to talk to you," Dr. Elisa said as she stuck her head in the door. She often had a tone of urgency in her voice, but today she sounded even more ruffled.

"What's up, Elisa?" Ringer asked.

"Did you hear about that twenty-six-weeker born on Friday? The one whose father is an OB intern?"

"Yeah, I heard something about that. How's the baby?"

"Not good. She has a unilateral grade 4 and a cerebellar bleed."

"Oh, shit."

"Yeah, but she's stable."

"What's the issue?"

"The dad asked me if he could take her home."

"You mean right now?" Ringer asked incredulously.

"No, not now," Dr. Elisa clarified. "He wanted to know if he had that right."

"Ohhh." Ringer exhaled slowly. "Got it. Thanks."

Dr. Elisa went back to the unit.

By Monday morning, Kelly and I had some basic information: Larissa was holding her own in the NICU but had a significant brain injury. The existing scientific literature on the outcome of children with injuries like hers, as I had discovered in the library, was dismal: universal movement impairment and a very high likelihood of diminished intelligence and cognitive function.

The problem with using medical literature to make predictions about an individual is that research studies describe the overall outcome of a large group of babies who are each unique. They might all have in common severe intraventricular hemorrhages and extreme prematurity, but some of the babies might also have had major infections that worsened their prognosis; others might have suffered respiratory injury; and still others might have had delayed growth during the pregnancy. The study described the average outcome—the median IQ score, the percentage of babies with severe motor impairment, the percentage of babies requiring a shunt to relieve high pressure in the brain—but a single baby is never the product of averages. Each baby is an individual, with a specific injury and a specific constellation of consequences of that injury.

Further complicating the picture for Larissa was that the extent of her intraventricular hemorrhage was uncertain, and it was not clear if she had bled in her cerebellum or not.

To better understand the extent of her injury, the doctors ordered a magnetic resonance imaging (MRI) study, which would be followed by a consultation with specialists in newborn brain injury.

Later that morning, two nurses freed the IV poles that supported Larissa's medications, intravenous fluid, and nutrition infusions, disconnected the oxygen tank from the wall, and bundled up her expanding medical chart. Then they headed out of the NICU with Larissa, one pushing the incubator, the other guiding the IV poles so they wouldn't stretch the tubes that fed through apertures into the incubator. Three floors below the NICU, they turned onto a bridge that crossed the narrow driveway separating Brigham and Women's from Children's Hospital Boston, and then they turned left down a ramp to get to the radiology department.

An older child needs to be sedated so he doesn't move during the thirty-to-forty-five-minute MRI exam, but a baby can usually be packed into the narrow MRI bed with sandbags surrounding his head to hold him still. During the scan, magnetic fields and radio waves jiggle protons in skin, bone, and blood, creating a stream of faint radio waves that paint extremely detailed pictures of brain tissue, the cerebrospinal spinal fluid that cushions the brain, and any blood that has escaped the vessels that are supposed to contain it.

During the nine months of gestation, the brain evolves from a ridge of cells into the most complex of human organs, but over this period, anatomic weaknesses related to the process of development leave the brain susceptible to injury.

The neural tube develops early in the second month of pregnancy; it bends and differentiates and then spreads to the sides. The results are the building blocks of the brain. The forebrain will form the cortex, the home of movement, thought, and the characteristics that distinguish us as human beings; the midbrain will become part of the brain stem, that primitive and hardy central stalk of the brain that connects to the spinal cord and controls basic and critical functions including breathing, heart rate, and temperature control. The hindbrain will go on to form the cerebellum, the sea-urchin-shaped structure at the back of the brain that coordinates movement and thought.

As the cortex develops, it pouches out to the sides and wraps around the ventricular system, which will produce and contain cerebrospinal fluid. The lateral ventricles, the third and fourth ventricle, and the subdural space form a circulation system for this clear fluid that bathes and cushions the brain and spinal cord, protecting it from injury. Through a sequence of foramina, or passages, fluid flows between the ventricles and around the brain and spinal cord.

Central to the cortex, enveloping the ventricles, are the precursor cells of white matter, the insulating cells that protect neurons—like rubber coating on a copper wire—and speed communication in the form of action potentials between one cell and another. Layered on top of these cells is the gray matter, the neurons that will form connections of enormous complexity.

Meanwhile, the vascular system is developing the four-columned foundation of a carotid artery on each side of it and a pair of vertebral arteries that feed the brain. At the top of the brain stem, the four arteries meet and feed a circular arterial system that then branches off to supply different parts of the cortex, brain stem, and cerebellum.

The muscular walls of the arteries will control blood flow and pressure in the brain by contracting and narrowing the blood vessels, like a hand pinching off a garden hose; blood flow and pressure beyond the contraction point diminishes. When fully developed, this system will allow blood to be preferentially sent to the brain in times of oxygen deprivation (the body prioritizes the most important organs—the brain, the heart, and the adrenal glands).

But early in development, this pressure-control system is immature, and blood flow in the brain resembles fluid going through a lead pipe rather than through a garden hose: pressure in equals pressure out.

In the fluid-filled lateral ventricles, one of which lies under each side of the cortex, the blood supply develops into a meshwork of tiny arteries called the germinal matrix. During the nine months of gestation, the germinal matrix evolves from a frondy network floating within the cerebrospinal fluid in the ventricle to a thin hardy strip of blood vessels that's firmly anchored to the ventricle's wall.

The combination in the immature brain of insufficient pressure control and weak germinal matrices floating inside the ventricles can be a lethal one if the baby is born too early. Blood pressure shifts wildly during a premature delivery, and the pressure within the brain can rise and fall sharply if uncontrolled by the arterial system, causing the fragile germinal matrix to burst and spew blood into the ventricular cavity. Named the germinal matrix because it gives rise to many of the hemisphere's neurons, this critical system is weak in the premature period only, an example of how biologic structures that are resistant to minor insults in maturity can nonetheless be particularly vulnerable to injury during development.

Think of an intraventricular hemorrhage as an explosion—one whose strength and consequences won't be known until later. The IVH can be mild—a little blood that leaks into the ventricular space that might not cause any long-term injury. These bleeds are given a 1 or a 2 on a scale of 1 through 4.

The blood from a more significant IVH fills the ventricle, clots, and then blocks the narrow passages that allow cerebrospinal fluid to circulate. With blood filling the ventricular space, and cerebrospinal fluid building up, pressure within one or both ventricles increases, which is when the real trouble begins. This is a grade 3 bleed.

The blood and pressure begin to strangulate the cells lining the ventricles—the cells that are supposed to develop into insulating white matter. The brain is compressing itself from the inside out. If bleeding from the tiny ruptured arteries continues, the pressure inside the ventricle builds, and the outward force in the ventricle can equal or exceed the arterial pressure supplying blood to the brain, cutting off the circulation and preventing surrounding brain tissue from receiving oxygen.

Once dead, these white-matter precursors don't regenerate. Over time, the dead cells liquefy, leaving clear spaces like the holes in Swiss cheese. The result is a slowdown in conduction from the cortex to the affected body parts—often arms and legs—so movement is slow or nonexistent, and the weakened signals from the cortex must fight the natural tendency of muscles to contract. This type of injury to the connection from brain to body is known as cerebral palsy.

Larissa's first ultrasound showed blood in both ventricles, more in the left than the right. But it also showed bleeding outside the ventricle on the left side—within the brain tissue itself.

This is referred to as a grade 4 bleed—the worst kind—because the bleeding doesn't just threaten injury to surrounding brain cells, it causes direct and permanent injury to brain cells in the vicinity of the bleeding.

The 1 through 4 grading system works well from the perspective of the consequences of the bleeding—grade 1 bleeds rarely have long-term consequences, and grade 4 bleeds always do.

Bleeds on the continuum from grade 1 to grade 3 represent blood in the ventricles. Grade 1 describes a bit of blood in the ventricle; grade 3 describes blood that clogs and causes dangerous swelling within the ventricular system. Some neurologists, however, believe that the grade 4 bleed results from a different mechanism because it is typically caused by bleeding from vessels other than the germinal matrix and because injury results from direct cell destruction, not from the pressure and irritation of the ventricle pushing outward. Ringer felt that, in Larissa's case, the injury was caused by trauma during her delivery.

Although the outcomes of babies with these severe hemorrhages have become somewhat better in recent years, when Larissa was born, the report of the definitive study of seventy-five babies with grade 4 bleeds like Larissa's was dismal: 59 percent of them died, and of the twenty-two children who survived to be evaluated later, 87 percent had major movement disabilities, and 68 percent had significant cognitive impairment.[1]

But brain injury in these babies isn't just a matter of grading. Why do some babies with a grade 3 IVH develop severe cerebral palsy while others with the same type of bleed exhibit only minor clumsiness? Part of the answer may be luck, but other factors clearly influence outcome. If a baby's blood pressure bounces up and down during the first hours or days of life, the brain may spend periods of time without enough oxygen, causing long-term damage to the brain cells. Infection can also influence the injury. The body's response to infection can lead to the dumping of chemicals called cytokines, which actually injure nearby cells—and if this occurs in the same area where bleeding has caused injury, the effect can snowball.

The unknown in Larissa's injury was the suggestion that bleeding had occurred in her cerebellum. When Larissa was born, there were few publications that described the long-term outcomes of cerebellar hemorrhage, but the importance of that part of the brain for the coordination of movement was well understood, and there was an emerging understanding among neurologists that the cerebellum also had an enormous role in learning, behavior, and information processing. A few years after she was born, a report on more than fifty children who had had cerebellar hemorrhages showed that half of them had movement impairment and a third had abnormal cognition, communication and behavioral abnormalities, and even autism.[2]

An hour after Larissa returned to the Brigham NICU, I put my white coat on over a pair of scrubs and walked across the bridge to Children's Hospital. I hadn't been in the hospital since I was a child, and coming in the back entrance, I found the place unfamiliar and disorienting.

"Hi, can you direct me to the MRI reading room?" I asked a secretary. My heart was racing, but I tried to affect the bored look of a tired resident.

"Through that door, right at the bottom of the steps," the secretary answered without really seeing me.

I followed her directions and pushed into a small, crowded darkened room stacked with video monitors. The smell of old coffee, potato chips, and poor ventilation wafted up. No one looked at me.

As my eyes adjusted to the light, one of the residents reading imaging studies called out, "Can I help you?" Neither he nor the three or four other residents and medical students who surrounded him glanced up.

"Yeah," I said nonchalantly, "I'm one of the OB residents from the Brigham, and one of our babies just had an MRI. Would you mind giving me a wet read?"

"Sure, no problem. Name?"

"Lowery. Baby Lowery."

"What's the story?"

"Twenty-six weeks. Now day of life four. Ultrasounds over the weekend showed bilateral bleeds and maybe a cerebellar hemorrhage as well."

There was a pause as the resident brought the study up on his dual-monitor display and scrolled through the hundreds of two-dimensional images of Larissa's brain sliced in all three planes by the MRI.

"Okay. Let's see. There is a left ventricular bleed, and also a moderate-size left parenchymal bleed," he said, describing the bleeding in the brain tissue adjacent to the left ventricle. "There is a little bit of blood in the right ventricle, but no evidence of bleeding. I think it may have come across from the left side. Both ventricles are normal size."

He turned toward the group sitting around him and peering over his shoulder. "You know, that's interesting. Here we have a hemorrhage that looks three to four days old and involves much of the left ventricle and the parenchyma around it, and yet the ventricle isn't dilated. That is unusual."

"What about the cerebellum?" I asked, trying to sound like this was almost an afterthought.

"Oh, yeah. Cerebellum looks pretty good. I see a small area here," he said, pointing. "Maybe there is a small bleed here. I'm not a hundred percent sure. I'll have to ask my attending, but if there is a bleed, it's a small one."

For the first time he turned toward me. "Anything else?"

"That's it," I said. "Thanks."

"Good luck," the resident said before clicking on the name of a new patient to bring up the images on his screens.

I walked back toward the Brigham to tell Kelly what I had heard.

Later that day, we entered the NICU to find Larissa completely surrounded by men and women in long white coats. The neurologists had arrived. Their attire was formal and their manner was somber. They were in the process of examining Larissa.

They measured her head circumference; they shone a flashlight into her eyes; they tested her primitive reflexes, pulling against her tiny arms and letting them spring back toward her to see if she retracted them equally, and then doing the same thing with her little legs.

They methodically flipped through her chart and wrote down her laboratory values. Then, in what seemed like single file, they walked to a computer monitor that allowed them to review the ultrasound images and MRI scans taken earlier that day. They spent some time arguing over whether there was a cerebellar hemorrhage or not, and they ended up agreeing that there probably was a small one, like the radiologist had said. Then the lead neurologist wrote a single sentence in the chart: *Neurology consult to follow.* And the line of doctors trooped out of the NICU and back to Children's Hospital so they could spend some time pondering what the future would hold for my daughter.

Already, four days after her birth, Larissa's brain injury was reasonably well defined, although it wasn't entirely clear if she had a cerebellar injury. If Larissa were an adult who'd had a similar injury as a result of a stroke or a traumatic event, the outcome would be predictable. You may remember the homunculus from high school science—the map of the body drawn on the brain. Scientists have mapped the regions of the brain that typically control specific parts of the body. Using this map and comparing it to the site of injury based on the ultrasound and MRI findings, neurologists could predict the extent of an injury.

But Larissa wasn't an adult—she wasn't even a full-term baby. Although Larissa had completed two-thirds of the length of a normal gestation, brain development occurs toward the end of pregnancy. At twenty-six weeks, the brain weighs only about 30 percent of what it will weigh at term, and because it is largely smooth, in contrast to the complex gyrations it will achieve later on in development, it has only about 20 percent of the surface area of what it will have at term. Whether Larissa's remaining brain growth would continue on a relatively normal trajectory—and the brain develops better in utero than it does in the NICU even under the best of circumstances—or be adversely influenced by the hemorrhage was entirely unknown.

The most important unknown—the question whose answer would determine Larissa's future—was the extent to which the unscathed brain tissue on the right would take up the tasks meant for the brain tissue on the left that had disappeared into an ugly glob of hemorrhage. Larissa's future would be determined by neuroplasticity.

The brain works by building and then maintaining or neglecting connections among neurons. For example, during development, a neuron that originates in the area near one of the ventricles can attach itself to cells near the brain surface and, using surrounding white-matter cells for guidance, build a scaffolding within the cell to support outward growth. Cells near the brain surface secrete chemicals called neurotrophic factors that serve as homing beacons for cell growth.

Neurons communicate with one another using action potentials: electrical charge builds up inside the cell and then, when a critical threshold is reached, travels down the length of the neuron's axon. The axon is the expressive part of the neuron, and its end lies in very close proximity to the receptive part of another neuron, called the dendrite; the axon and dendrite are separated by a very small space, called the synapse.

When an action potential travels down an axon and reaches the end, the axon releases a type of chemical called a neurotransmitter into the synapse. The neurotransmitter fits, like a key into a lock, in receptors on the dendrite of the neighboring neuron. The combination of neurotransmitter and receptor opens gates in the cell wall, and charged molecules flood in and create an action potential in the neighboring cell.

I am simplifying this next part of the process dramatically, but imagine that an action potential has fired in a neuron in the part of my brain that handles expressive language. From one neuron to another, action potential to neurotransmitter to action potential, the signal is relayed to my left motor cortex, the region of the cortex that handles voluntary movements. Here, the signal is passed to a neuron that extends all the way down the spinal cord, crossing from the left to the right side at the base of my skull. Then the signal is transmitted to yet another neuron, one that stretches the entire length of my right arm, all the way to a specific muscle in my right ring finger. At that muscle fiber, the axon releases acetylcholine, a neurotransmitter that doesn't cross to another neuron's dendrite but instead triggers the firing of the muscle that makes my finger contract and hit the period key on my keyboard that's needed to end this sentence.

When I was learning to type, the process was clumsy and slow. Over time I got faster and increasingly accurate. In my brain, there was no proliferation of new neurons, but rather an increase in the density of the receptors on the neurons involved in this circuit. With each new receptor added, the amount of neurotransmitter needed to trigger the action potential diminished. This is how learning occurs.

The homunculus map that links a specific part of the body and its function to a single, isolated region of the brain originally came from research that indicated that the map was an immutable system—that is, once a function was localized to a particular part of the brain, the location of that function would never change. Every medical student learns about Broca's area, a section of the left frontal lobe associated with the motor control of speech. The function of the area was described by the physician Paul Broca in 1861 after a patient who had lost the ability to speak was found on autopsy to have a syphilitic lesion in the left frontal lobe. In Broca's patient, the injured frontal lobe couldn't accomplish the "thought" that initiated speech—the connection between the brain and the muscles of the mouth and tongue worked just fine, but no signal initiated speech. The fixed nature of the homunculus—locationism theory—became dogma, and for nearly a century, any suggestion that the fixed system wasn't all that fixed was ignored.

In the 1960s and 1970s, research started to chip away at this orthodoxy. Experiments on monkeys showed that when a brain was rewired—when the optic nerve was surgically directed toward the auditory cortex instead of the visual cortex, for example, or when nerves from different fingers were surgically swapped on their way to the brain—the change was accommodated. The auditory cortex, typically the area that processes sound, learned to process visual input. The brain relearned the location of the fingers to allow for normal hand function.

Clinical experience was undermining locationism theory. In his fascinating 2007 book on neuroplasticity *The Brain That Changes Itself*, the psychiatrist Norman Doidge described the prolonged and intensive at-home rehabilitation of Pedro Bach-y-Rita, a man who had a devastating stroke that left him paralyzed. Instead of allowing his father to waste away in bed, Bach-y-Rita's son worked with him for hours each day. He re-taught his father to crawl, kneel, and finally walk, and, by repeatedly practicing crude motions with his hands, arms, and fingers, his father slowly regained normal function. Pedro Bach-y-Rita's son, Paul, happened to be a leading scientist, and he was as responsible as anyone for reviving the science of neuroplasticity in the 1970s. When Pedro Bach-y-Rita finally died, of a heart attack while mountain climbing in Colombia, his autopsy showed that 97 percent of the connections between his cortex and his spinal cord had been destroyed—he had regained complete function of his arms and legs by teaching the remaining few neurons to take over the tasks of their destroyed neighbors.

Back at Children's Hospital, the neurologists reviewed Larissa's MRI and then sat down in a conference room to discuss their findings. The medical literature predicted awful outcomes for children with grade 4 intraventricular hemorrhages like the one Larissa had. But there was anecdotal evidence suggesting that this wasn't always the case.

Adre du Plessis, a brilliant and profane neurologist originally from South Africa, had a particular interest in premature-infant brain injury, and he'd been collecting cases for a report on grade 4 intraventricular hemorrhages. His research showed that kids with unilateral grade 4 hemorrhages often weren't as devastated as those with the more common bilateral type of hemorrhage. He thought that the reports his colleagues were reading were crude, and the awful outcomes might be influenced by the larger number of bilateral hemorrhages in the studies. One of his research fellows had analyzed the group with unilateral bleeds and found that, as he put it, "The numbers were a hell of a lot better" than what had been described. "Functionally," he said, "these kids were in the game."

For Larissa, the goal would not be to reassign neurons to do the work previously done by her destroyed brain cells. Rather, she would depend on that plasticity from the very start, relying on completely new organizations and neural connections to operate her right arm and right leg and do the complex and subtle tasks involved in cognition.

The neurologists at Children's argued a bit over how to frame their prediction. They knew we were looking for certainty, and they knew there was none to be had. Even so, as du Plessis liked to say,

"There's no gold medal for identifying things you can do nothing about."

Eventually the troop of neurologists in white coats came back to the Brigham NICU. The attending led the way to Larissa's bedside, and they all stood in a semicircle while he wrote a note in the chart.

As doctors tend to do, the attending neurologist briefly restated her injury and noted the imaging that had been done.

Then he wrote two lines under the heading of Prognosis.

Chance of movement impairment on the right side is 100%—degree of impairment is difficult to predict.

Chance of normal cognitive function (IQ greater than 70): 50%.

CHAPTER 5

Injury, and What Follows

Review books and index cards piled around him, Jason Carmel was deep into the rite of passage familiar to every medical student—studying for Step 1 of the United States Medical Licensing Examination. The anatomy of the inner ear and the biochemistry of the renal collecting system were swirling around in his head when his cell phone rang.

His twin brother, David, was traveling with friends in Puerto Vallarta, Mexico. He'd been playing soccer on the beach when the group decided to go swimming to cool off. David ran into the surf until the water got too deep, and then he dove in, ramming headfirst into a sandbar that lay hidden beneath the waves.

Information was sketchy at first: He had fallen and hit his head. He had fallen and couldn't move his legs. Throughout the day, phone calls flew back and forth between David's friends, who were with him, and David's family on the East Coast, particularly from the twins' father, a neurosurgeon. At the end of the day, David was airlifted to San Diego to undergo emergency surgery to stabilize his spine fracture.

The next morning, Jason Carmel sat next to his father on a cross-country flight with his anatomy atlas in his lap, trying to understand the prognosis of his brother's injury. "I was a second-year medical student," Jason remembered. "I knew it was serious, but I didn't understand what *serious* meant."

In the intensive care unit, Jason began to comprehend the consequences of his brother's injury. David had suffered a burst fracture of the sixth cervical vertebra, and the bone fragments had injured his spinal cord. Because the neurons in the brain control movement through the corticospinal tract; that tract had been injured, so communication that orchestrates movement at the site of injury and below had largely ceased. David was paralyzed below the chest; he could lift his arms, but he had only minimal movement in his hands. When David arrived in San Diego, surgeons had removed the shattered vertebra and installed screws to support his spine. Over the next week X-rays, MRIs, and an additional surgery followed in rapid succession.

"I remember sleeping in a motel near the hospital, and it didn't feel right," Jason recalled. "I was lying down in a soft bed while David was immobilized in a hard collar in the ICU."

Eventually, David was discharged from the ICU and flown home to begin inpatient rehabilitation at Mount Sinai Medical Center in New York.

Jason Carmel finished studying for his exam, splitting his time between his apartment and the medical library at Mount Sinai, which was downstairs from his brother's room. Then he started his third year of medical school at Columbia and began the series of required rotations that teach students the basics of a wide variety of medical specialties—from internal medicine to surgery, psychiatry to obstetrics—and help them decide what they want to specialize in when they graduate.

But he found himself unable to focus. "I was directionless that year," Jason remembered. "The normal energy that people apply to figuring out their lives, I lost in my distraction with David's injury. I didn't know what I wanted to do."

Jason and David Carmel, and their older brother, Jonathan, had grown up surrounded by doctors; their father was a pediatric neurosurgeon, and both of their grandfathers were physicians. Jason knew he wanted to be a doctor, but unlike many of his classmates, who one day realized that they were destined to be gastroenterologists (or otolaryngologists, or family practice doctors), Jason Carmel had no epiphany that revealed his future specialty.

One afternoon Jason was visiting David when Wise Young, an eminent researcher in spinal cord injury and a longtime acquaintance of their dad, stopped by the drab room at Mount Sinai. "Here was this bearded Asian guy," Jason recalled, "who came into the hospital room and sat down and held David's hand and talked about spinal cord injury and what he was trying to do in his lab. He was warm and caring and showed enormous empathy."

Young was also brilliant and successful. He had gone to medical school and trained for two years as a neurosurgeon before leaving the OR for the lab. Listening from across the room as Young talked about his research in understanding and treating spinal cord injury, Jason found himself transfixed. "It seemed there was energy and promise in this field of research, and I suddenly felt like I wanted to participate in that."

Jason Carmel took a year off from medical school to do research and convinced Young to let him work in his Rutgers University lab. New to basic science, Carmel hurried to learn pipetting and sample preparation from graduate students and research assistants in the lab.

That summer, the lab was enamored of a new technology that had the remarkable capacity to evaluate the function of dozens of genes simultaneously.

In one experiment, Carmel simulated human spinal cord injury in a rat: he dropped a half-ounce metal rod from a height of precisely one inch onto the rat's exposed spinal cord, and then he used the GeneChip to see what happened to the function of spinal cord genes below the site of injury.

There was so much data about so many genes that Carmel and his colleagues had trouble figuring out what the results meant; they didn't know what to do with the information other than publish a long paper. Although the practical benefit of all that knowledge seemed elusive, the research was a starting point for the young investigators.

Carmel's year in the lab flew by, and his team began to focus on a specific question: Why can injured neurons regenerate in some parts of the body but not in others? They teamed up with another lab that had concocted a compound that allowed neurons to grow anywhere, even in parts of the body where they were usually dormant. The researchers applied the compound to neurons, then used the GeneChips to see what genes were expressed in the neurons after the growth treatment; they found a number of genes had been turned on by the treatment. (All genes are present in all cells all the time, but only a fraction of genes are functional, or expressed.) This new research had the potential for useful therapy: in areas where injured neurons did not heal themselves, such as the spinal cord, the growth compound could turn on genes that were not typically expressed and possibly lead to nerve regeneration. To Carmel, working in the lab felt like pushing the bounds of what was possible in a way that clinical rotations had not.

Three months after entering Mount Sinai, David was discharged, and he settled into a routine as he adapted to his post-injury life. In time, he learned to be largely independent. An aide came in each morning to help him bathe and dress, but once he was in his wheelchair, he could navigate the world on his own.

At the time he was injured, he'd been about to start at the Stanford Graduate School of Business; a year after his injury, he was ready to get on with his life, and David moved out to the West Coast.

In Wise Young's Rutgers lab, a year of research turned into two as Jason's interest in spinal cord injury deepened and focused. "I didn't have any illusions that I would be the one that got my brother more sensation or more movement," Carmel said, but he acknowledged that the symmetry between David's injury and his research path wasn't accidental.

Jason learned techniques and ran statistics on the GeneChip results for the lab's research on the genetics of spinal cord injury, but he found himself inexorably drawn to research directed at protecting the spinal cord from injury or, better yet, regenerating it after injury.

Jason found his social calendar disturbingly empty after David left New York. He finally got around to calling the daughter of a family friend who had started medical school at Columbia a few years after Jason. It was meant to be: in 2005 Amanda and Jason were married.

Plasticity is the ability to change, and we are all plastic to the extent that we learn from experiences and adapt. Neuroplasticity is the adaptation of neurons and the connections they make with one another in response to experience.

In his 1890 classic *The Principles of Psychology,* William James proposed the theory of neuroplasticity. In subsequent decades, a few researchers published papers providing evidence for neuroplasticity. But in the 1930s, the neurosurgeon Wilder Penfield used mild electric stimulation to identify the location in the brain where intractable seizure activity originated. In the process, he created detailed anatomic maps that linked specific locations in the brain to motion and sensation in specific parts of the body. The combined information of these maps was published as the homunculus, and it remains essentially uncorrected to the present day. The illustration fed a popular fascination with the elegant and logical layout of the brain and the concept that specific cortical real estate was assigned to specific anatomic structures. An age of locationism set in and lasted about fifty years; the earlier literature was largely ignored, and researchers came to the conclusion that the homunculus was not plastic and that brain injury led to permanent loss of function in the part of the body that had been unlucky enough to be wired to the injured section of the cortex.

It became dogma that after childhood, the structure of the neurons and white-matter cells that made up the brain—the brain stem, the midbrain, and the cortex—was essentially fixed. The two exceptions were the areas that housed memories—the hippocampus and the dentate gyrus—which continued to grow new neurons into adulthood.

Research into neuroplasticity began with the map. In the 1970s and 1980s, it was discovered that if the input going to a part of the map was removed—for example, if information from the middle finger stopped arriving—the area of the cortex that had been assigned to the middle finger did not lie dormant but rather began to participate in the coordination of other structures, typically nearby structures, such as, in this case, the second and fourth fingers. Conversely, if a tiny stroke obliterated the area of the cortex devoted to the middle finger, adjacent areas rapidly took up control of the middle finger and restored its function. This research was expanded through experiments using fine electrodes inserted into precisely mapped spots on the cortex and showed that the area of cortex dedicated to a specific anatomic location would increase if that part of the anatomy was used and stimulated extensively and became important from the perspective of the animal being studied.

The ability of the brain to reorganize confounded scientists' efforts to trick it. When neuroplasticity pioneer Michael Merzenich cut a peripheral nerve in a monkey and sewed it to an adjacent nerve, he expected the experiment to utterly confuse the monkey's brain. But when he mapped the monkey's cortex after a period of adjustment, he found the monkey's brain had rectified the map, accommodating the nerve change by reorganizing inputs.

Plasticity was enhanced when the desired movement was rewarded, as is often the case in psychological experiments. The behavior being studied developed more quickly and the area of the brain devoted to it got larger when rewards were given than when they weren't.

Maybe the most innovative neuroplasticity research was done by Paul Bach-y-Rita, who developed ingenious devices that leveraged neuroplasticity to overcome catastrophic impairments. For one woman, who'd been incapacitated by unremitting vertigo after the collapse of her vestibular system—the balance equilibrium system in the inner ear—Bach-y-Rita developed an external balancing device that utilized tiny shocks to the woman's tongue to provide equilibrium input. Over a period of months, the device allowed the woman to regain her balance, proving that the brain can receive information about balance from the tongue as well as the vestibular system. Eventually, she even outgrew the need for the device.

Another of Bach-y-Rita's inventions was a camera that converted its video input into mechanical vibrations using stimulators connected to a chair. Over time, a research subject born blind gained the capacity to "see" by sitting in the chair and feeling the inputs against his skin; he was able to distinguish forms and pictures placed in front of the camera by way of the vibrations. Although Bach-y-Rita did not map the part of the brain that was processing the input from his chair, he proved that even in someone who had never seen, the brain could process images from an unlikely source of sensory stimulation—the skin.

It is interesting to note that the research on the development of vision in kittens, one of the key series of experiments that explained neuroplasticity, was partly wrong. David Hubel and Torsten Wiesel showed that if one of a kitten's eyelids was sewed closed soon after birth, the part of the visual cortex usually assigned to that eye learned to process visual information from the opened eye, and it continued to do so even when the sutures were later removed from the closed eyelid. The previously closed eye did not regain sight, and the researchers believed this was evidence that there was a critical period during which plasticity exists, and that, at the end of this period, plasticity ends (or is greatly diminished). In fact, this experiment may better represent evidence of learned nonuse. For the kitten, it was just plain easier to use the eye it had always used than to learn how to see out of the other eye.

This same principle applies to motor connections that are disrupted early in life. An animal—or child—will use the limb that works the best; sparse connections can be forced to develop only if the subject is constantly encouraged to use the injured or affected limb.

I met Sarah Habib in her comfortable New Hampshire home when she was five years old. Sarah was born fifteen weeks early, and her birth had all of the drama of Larissa's: Sarah's mom, Kim, had felt sick all weekend but chalked it up to a stomach virus. Monday morning Sarah's father, David, left for Memphis on a business trip, and the next day Kim's water broke and she rapidly went into labor. Sarah arrived feet first at a community hospital, and she was placed into an incubator and put in an ambulance that got her to the NICU at Tufts Medical Center in Boston in forty-five minutes.

As rocky as Larissa's initial days in the NICU were, Sarah's were even bumpier. She had hemorrhages on both sides of her brain, and the doctors struggled to move oxygen from the ventilator through her lungs and into her bloodstream. They turned up the oxygen setting on the ventilator, knowing it might cause blindness but realizing that they had no alternative. Finally they brought in the ventilator of last resort, a machine often referred to as the jet because of the sound it makes; it got oxygen into Sarah's lungs but made Sarah vibrate with each of the hundreds of tiny puffs of air a minute it sent into her lungs.

One night after Kim and David had made the eighty-minute drive home to New Hampshire, the phone rang and the doctor on the other end of the line told them in a grave voice that they had better return to Tufts. Weary beyond speaking, they drove the abandoned highway back into Boston, where, afraid that Kim and David might arrive too late, the nurses had taken some photos of Sarah, just in case.

It was the whole nightmare: intraventricular hemorrhages, lung failure, and now trouble maintaining Sarah's blood pressure. Her doctors came to Kim and David and suggested that they sign a do-not-resuscitate order so that in the event that Sarah's heart stopped beating, the pediatricians wouldn't be compelled to perform heroic measures on this baby who seemed so fragile and whose outcome was so uncertain. Kim and David signed the order.

But over the next several days, Kim and David marveled at how their tiny newborn—their only child—hit obstacle after obstacle, complication after complication, and yet managed to survive.

"We decided that if she was going to fight, then we would fight with her," Kim said. The next day they rescinded the DNR order.

Meanwhile, David was online looking for hope. He tracked down doctors around the world who had taken care of premature newborns with severe neurologic injury and had seen miracles. He read research on novel ways of encouraging development. "I was e-mailing with a doctor in the UK about therapies that could rewire the brain," he recalled. "They had done some research, and maybe it was really new, but it made me optimistic."

Their Tufts neurologist was circumspect. "Well, you know," was all he would say when David mentioned his far-flung communications.

David wasn't ready to accept less than normal. "I did not believe it," he said. "I still believed that given the right opportunities, this kid could do whatever she wanted to in life."

Ever on the lookout for hopeful signs, Kim found one on the wall of the subway station below the hospital. LOVE CONQUERS ALL, it said—it was an advertisement for something, but to Kim it was a message aimed at her and Sarah.

CHAPTER 6

Whose Choice?

Wrapped in her blanket inside the warm and humidified incubator, Larissa was beginning to grow. Her demands on the NICU staff diminished each day. Whereas she had once required the ventilator, now she needed only a whiff of oxygen under her nose. Her blood count was stable, and the antibiotics that were being infused to ward off an infection seemed like overkill. Though, to be sure, she still needed formula dripped through a tube into her stomach and the warmth and protection that the incubator provided.

However, even as Kelly and I and the nurses and doctors stared through the Plexiglas at this thriving newborn, we couldn't avoid worrying about the devastation to the left of the midline within her skull.

Here was a child who had a dubious prognosis—the risk of devastating disability was still high—but who was completely stable.

Stephen Wall and John Colin Partridge, pediatricians at Northwestern and the University of California at San Francisco, studied the records of 165 newborns who had died in the San Francisco NICU over a three-year period. Of these babies, 121 had died after doctors withdrew life-sustaining care or decided not to provide the additional care needed to prolong life. The most common reason for limiting or withdrawing life support was the physician's belief that death was inevitable and that additional treatment was futile. Doctors who stopped care also expressed concern that the baby would suffer as a consequence of futile treatment.

What is remarkable was that in each of these 121 cases, the parents and the physicians agreed on the decision to stop treatment. In not a single case did the parents and the doctors disagree.[1] Parents and doctors make these decisions every day in NICUs all over the United States. And although these decisions are accompanied by great sadness, there is rarely any controversy.

Usually, these choices are simplified because babies tend to make the big decisions themselves. After some uncertainty, they get progressively better: the ventilator is replaced by simple oxygen supplementation, feeding is begun, and the overall care requirements diminish. Or one complication follows another—intracranial bleeding is diagnosed, increasing doses of medication are needed to maintain blood pressure, and the oxygen requirements go up, not down.

The fact is that parents and physicians share a common desire to do everything medically reasonable to save the life of a premature baby. These are not tepid resuscitations or halfhearted attempts but all-out efforts with extraordinary resources marshaled to save the baby's life. Parents typically see committed teams of doctors and nurses working around the clock, and they derive confidence and reassurance from the compassion and commitment that motivate their baby's caregivers.

It may not be surprising, then, that when the doctors realize that death is inevitable and approach the parents with this devastating news—and it is almost always the doctor who initiates the discussion about ending care—parents are receptive to the doctors' suggestions.

The withdrawal of life-sustaining care has been a part of medicine for centuries. With each advance in the care of newborns, both full-term and preterm, physicians and parents found themselves asking, as child psychiatrist Leon Eisenberg put it, in a 1972 article, "At long last, we are beginning to ask, not can it be done but should it be done."[2] But in an era of relative medical paternalism, the conversation was held in hushed tones.

The public discussion on this topic began in 1973 with the publication in the *New England Journal of Medicine* of a lucid and thoughtful article, "Moral and Ethical Dilemmas in the Special-Care Nursery."[3] It was written by two Yale pediatricians who looked at the causes of death of the 299 babies that died in the Yale NICU over a two-and-a-half-year period. In 14 percent of those deaths—most of them children with congenital anomalies—care was withdrawn with the agreement of parents and caregivers.

Most of the injuries and illnesses described in the article were cases where the majority of reasonable contemporary pediatricians and parents would agree that further care was only delaying death: the child with respiratory distress syndrome on a ventilator requiring a lot of oxygen who developed heart disease and seemed to be suffocating from oxygen deprivation; the child with multiple abnormalities in the structure of the pelvis and pelvic organs, severe spina bifida, and, as a consequence, a buildup of pressure around the brain.

The 1973 article described how, in making this decision, physicians and parents considered the quality of life for these children, as well as the potential for physical and emotional suffering. But parents and physicians also made the decision to withdraw support for children who would likely survive but would suffer severe disabilities; they considered the toll that would take on their families and on society, as well as the financial burden that their care would cause. Based on these deliberations, care was withheld from babies with conditions that, even in 1973, were survivable, including an infant with Down syndrome and a correctable malformation of the intestines who did not receive surgery and was allowed to die.

In most of these cases, physicians and parents agreed that death was the preferable outcome. In the few instances in which the physicians thought the prognosis was dismal but the parents wished life-sustaining treatment to continue, it did.

Although it seems inconceivable, there is no evidence of public disagreement with the doctors, no case in which the parents asked that care be stopped and the physicians insisted that it continue. This may reflect those times—the patient-physician balance of power was even more skewed than it is today—or a prejudiced belief that disability, physical or cognitive, was incompatible with a high quality of life. Either way, in these early years of neonatology, parents and their doctors had relatively few constraints on their freedom to make life-and-death decisions for newborns like Larissa whose prognosis was uncertain.

"We believe," wrote Raymond Duff, lead author of the 1973 article, "the burdens of decision making must be borne by families and their professional advisers because they are most familiar with the respective situations."

Some years after Larissa was born, I admitted a woman to our Labor and Delivery service whose membranes had ruptured at twenty-three weeks and four days. The patient was highly educated, as was her husband, and they were accompanied by an adorable three-year-old daughter, who charmed our staff. The woman was not in labor, and so we counseled her that there was a reasonable chance that she would remain pregnant for a few days at least, maybe longer.

"I have to warn you," I remember telling her, "that most women who rupture their membranes do go into labor within forty-eight hours. But the longer you go without labor, the higher the chance that you'll remain pregnant for an extended period of time."

As a matter of routine, I called the NICU for a consult, and one of the neonatologists came and spoke at length with my patient.

A couple of hours later, I returned.

"The neonatologist said that if my baby is born today, survival is about fifty percent," she said, "but she is almost certain to be injured. And if I stay pregnant for another three days, and I get two full doses of steroids, survival will increase to maybe sixty-five percent."

"That's good," I ventured.

"What are my options?" she asked. A severe tone had entered her voice.

"What do you mean?" I asked her. I had prepared to be hopeful.

"Can I end the pregnancy?" she asked. "Can I have an abortion? Or is it too late?"

I was taken aback. I explained that the legal limit for abortion in Massachusetts was twenty-four weeks, and so in theory she had three days. "It may be hard to find a doctor in Massachusetts who will do an abortion for you so close to the legal limit," I told her.

"And besides, your baby could do really well, or you could stay pregnant for weeks . . ."

"Or my baby could be born tonight and end up with catastrophic brain damage. The other doctor I spoke to mentioned brain damage. Isn't that true?"

I had to admit it was.

Our discussion continued for close to an hour. When I left the room I went to the nurses' station. "She wants an abortion," I said.

The nurses looked at me like I was crazy. "She wants what?" one of them said, incredulous.

They took it as a personal affront, this highly trained group of professionals who were preparing to do everything necessary to save the life of a severely premature baby, that my patient would decline their care and seek an abortion.

Grudgingly, we disconnected the woman's IV; she changed out of her hospital gown and got back into her street clothes, walked out of the hospital, and took a taxi across town, where one of my colleagues at another hospital performed her abortion.

Most of my colleagues, as well as the nurses who were personally wounded by this woman's rejection of their skills and good intentions, would characterize themselves as strong advocates of abortion rights. Yet it felt wrong to these providers of women's health care that my patient had ended a pregnancy that might have resulted in a lovely sister for her delightful three-year-old because she didn't want to risk the chance that her child would be disabled. They felt she had made the wrong decision, but no one called the hospital's lawyer or phoned the attorney general; they questioned her decision, not her right to make it.

My patient chose abortion at twenty-three weeks and four days, and yet, had she gone into labor and delivered, her baby could have received all the resources our NICU had to offer. Because these decisions hinge on mere days, it is tempting to group together the abortion debate and arguments over who should make decisions about the care of sick or premature newborns, but the legal and medical traditions of the two situations have entirely different origins. Society has debated the morality of abortion for centuries, but at the center of the argument is a disagreement regarding the balance between a woman's rights and the rights, or lack thereof, of the fetus. Abortion in the United States has always been either outlawed or restricted; it has never been a private medical decision made legally between a woman and her doctor. My patient considered her options within the highly structured system of American abortion laws.

In contrast, the legal framework for making decisions about which babies get the overwhelming resources of a modern NICU and which babies are allowed to die is less mature and hence more tenuous. The decisions in the nursery involve a baby who has rights and who is entitled to full legal protection. The debate isn't about whether the child has rights (it does), but rather who is empowered to make decisions for the child.

Until the 1970s, these decisions were the respected province of parents in consultation with physicians, in part because societal prejudice against the disabled aligned conveniently with the presumption that parents would make appropriate decisions in the best interest of their child.

Then conservatives, having lost the abortion debate in *Roe v. Wade* and fearing the breakdown of society, began to look for ways to shore up social structures. They found it in the case of Baby Doe, a child with Down syndrome born in Bloomington, Indiana, in April of 1982. Soon after birth, the child was diagnosed with a tracheoesophageal fistula, a condition in which the esophagus fails to communicate with the stomach and instead links sideways to the trachea. A relatively simple surgery restores the channel so sustenance can reach the stomach and blocks the fistula that would otherwise allow food and liquid to leak into the lungs. Without the surgery, the child would die of malnutrition; with surgery, survival was nearly guaranteed.

The parents refused to sign a consent form permitting surgery. Outraged, the child's pediatricians and the administrators at the hospital where the child was born went to court to obtain the needed permission. The Indiana Supreme Court sided with the parents, and while the hospital was in the process of appealing to the U.S. Supreme Court, details were leaked to the media, and the case gained national attention. Six days after he was born, with several families declaring that they would be willing to adopt the child, and with the high court appeal pending, Baby Doe died of malnutrition.

Outraged by the court's decision, President Ronald Reagan ordered the Department of Health and Human Services to protect future Baby Does. The political charge was led by the surgeon general at the time, Dr. C. Everett Koop, a pediatric surgeon and founder of one of the nation's first NICUs. The resulting regulations, known as the Baby Doe rules, undermined the premise that parents, with the guidance of caregivers, were the ones best suited to make decisions for their children.

The month after Baby Doe died, government health administrators informed hospitals that they could lose federal funding if they withheld treatment or nourishment from handicapped children. Another set of regulations that followed required hospitals to post large signs stating DISCRIMINATORY FAILURE TO FEED OR CARE FOR HANDICAPPED INFANTS IN THIS FACILITY IS PROHIBITED BY FEDERAL LAW. The posters provided the number of a twenty-four-hour, toll-free, anonymous hotline for people to report violations.

NICU physicians began looking over their shoulders, fearful that anyone who disagreed with one of their decisions might trigger the arrival of groups of federal investigators that came to be known as the God squads. Without warning, a group composed of federal civil rights investigators and consulting doctors would show up at the hospital and demand access to the child, the child's caregivers, and all the related documentation. Whether or not the investigators found a violation, their visits were chilling.

Angered by the Baby Doe rules, which not only disempowered parents but also dictated how doctors ought to practice medicine, physicians fought back. They argued that the government had no business usurping the decision-making rights of parents and their children's doctors, particularly with rules so broad and overarching that they forced doctors to continue life-sustaining care on terminally ill children regardless of the pain and suffering this futile care caused. In a lawsuit decided by the U.S. Supreme Court, the Baby Doe rules were rescinded and the God squad posters came down. But by that point, the era in which parents and caregivers could reach a decision about what was in their child's best interests without considering the legal implications of their actions was over.

A second set of Baby Doe rules was enacted in 1985, this time by Congress, imposing similar legal requirements on the care of premature and sick newborns, although without the God squads or combative posters. Language in the new rules included the phrase *reasonable medical judgment,* which assuaged the ire of the American Academy of Pediatrics, and a détente set in. When Congress passed the Born-Alive Infants Protection Act in 2002, pediatricians were quick to point out that the new law didn't change the status quo. The God squad period had frightened pediatricians into accepting a set of legal guidelines that governed decisions made about sick or premature newborns, but because federal officials weren't aggressively imposing a whistle-blower environment inside hospitals, the new rules didn't spark a full-fledged revolt.

The environment in the newborn nursery shifted from one that granted caregivers and parents the autonomy to make decisions for their children to one that was open to controversy, and, inevitably, disagreements between parents and doctors began to end up in court.

The best-known contemporary disagreement involved Sidney Miller, now an adult who cannot walk, speak, sit up, or care for herself. She is blind in one eye and has minimal vision in the other. She has cerebral palsy with spastic quadriplegia, a seizure disorder, and hydrocephalus requiring a shunt to drain fluid from around her brain into her abdomen. She has mental retardation with a cognitive capacity equivalent to a six-year-old's. Her mother, Karla, provides around-the-clock care for her routine needs—diaper changes, feeding, and dressing—as well as her unique medical ones.[4]

Karla arrived at Woman's Hospital of Texas on August 17, 1990, with symptoms of preterm labor. An ultrasound estimated that Karla's fetus weighed 629 grams—just over one pound—and was approximately twenty-three weeks gestational age. Soon thereafter, Karla's obstetrician, Mark Jacobs, diagnosed chorioamnionitis—an infection in the uterus—and recommended delivery. The alternative would put Karla's life and the life of her fetus at risk.

Jacobs and the neonatologist on call, Donald Kelley, counseled Karla and her husband that their daughter probably would not be born alive, and if by chance she was, she would likely suffer from the disabilities that plagued severely premature newborns: brain hemorrhage, blindness, and lung disease.

The two doctors asked Karla and Mark Miller to decide whether or not they wanted their child aggressively resuscitated at birth, and after weighing the information they had received, the Millers told the physicians they wanted "no heroic measures" taken when their daughter was born. Dr. Kelley documented the Millers' request in Karla's chart. Mark Miller left the hospital to make funeral arrangements for the daughter who would soon be born and would then die.

The decision not to fully resuscitate the baby alarmed hospital staff, who conferred in a series of rushed meetings and then informed Mark Miller on his return that that hospital policy as well as state and federal law required the staff to resuscitate any infant born weighing more than five hundred grams. When Mark requested a copy of the policy, he was told it was an unwritten one, and when he asked how he could prevent a resuscitation from happening, he was told he would have to remove his wife from the hospital. Given Karla's worsening infection, Dr. Jacobs made it clear that moving Karla was not a safe option.

Karla's water broke that evening, labor was induced, and at eleven thirty, she delivered Sidney, at twenty-three weeks and one day of gestation and weighing 615 grams.

Although Mark Miller had refused to consent to his daughter's resuscitation, the neonatologist on call at the time of Sidney's birth, Dr. Eduardo Otero, evaluated Sidney and decided that the baby was alive and had a reasonable chance of living. He resuscitated Sidney, intubated her, and put her on a ventilator.

Sidney's first days in the NICU were stable, but she quickly developed a significant intraventricular hemorrhage that resulted in the devastating disabilities that remain with her today.

The Millers sued the hospital and its parent company, Hospital Corporation of America, claiming battery because Sidney was resuscitated without the consent of her parents, and negligence. After a one-month trial, the jury awarded the Millers $60.4 million in actual and punitive damages. But the Texas Court of Appeals overturned the verdict. In the opinion of that court, "A physician, who is confronted with emergent circumstances and provides life-sustaining treatment to a minor child, is not liable for not first obtaining consent from the parents. . . . [This is] an exception to the general rule that a physician commits a battery by providing medical treatment without consent." The court deemed Sidney's birth an emergent situation because her condition could not be evaluated prior to birth.

Although the Hospital Corporation of America's defense rested largely on compliance with the Baby Doe law, the court seemed more interested in providing physicians with legal cover when they made "emergency" decisions.

Bioethicist George Annas's interpretation of the Texas ruling was that "the court implies that life is always preferable to death for a newborn and thus could be interpreted in the future to support the neonatologist who always resuscitates newborns, no matter how premature or how unlikely their survival may be without severe disabilities."[5]

Whether based on Texas law or the Baby Doe rules, this dispensation to resuscitate first and ask questions later seems to be backed by essentially all legal venues. It is worth noting that these legal principles are not medical paternalism wrapped up in legalese. In the few cases where parents wanted more aggressive treatment than physicians thought was reasonable, the courts sided with the parents and affirmed the principle that life is always preferable.

A 1994 case pitted the mother of Stephanie Keene, who became known as Baby K, against Fairfax Hospital in Virginia, where the child was born. The infant was anencephalic—a lethal syndrome in which the fetus develops without most head and brain structures. The condition was diagnosed by ultrasound during the pregnancy, but her mother insisted on full resuscitative efforts even though death was certain. The pediatricians, backed by the hospital, argued that such care would be futile and, since many resuscitation procedures are painful, inhumane because the child would certainly die. The child was intubated at birth, and moved to a nursing facility at six months of age. Even though all agreed that survival for much longer was not possible, a court of appeals ruled that the child was entitled to continued treatment if the mother requested it.[6] On and off a ventilator, and without ever exhibiting any behavior that required more than her brain stem, Stephanie survived until age three and then died at the hospital where she'd been born.

Other courts have come to the same conclusion as the appeals court in the Baby K case, even when survival isn't just unlikely but fundamentally impossible. Shemika Burks arrived at St. Joseph's Hospital in Milwaukee complaining of contractions about twenty-two weeks into her pregnancy. An hour later, she delivered her daughter, Comelethaa, who weighed two hundred grams (seven ounces). It is the American Academy of Pediatrics' position that pediatricians should not resuscitate babies who weigh less than four hundred grams or are less than twenty-three weeks gestational age.[7] Shemika Burks asked that the hospital staff do what they could to save Comelethaa, but the neonatologist refused and explained that even though the child had a heartbeat at delivery and moved her chest as if to breathe, she could not possibly survive, and so it would be cruel to use heroic measures. Hospital staff wrapped Comelethaa in a blanket to keep her warm, and Shemika Burks held her until she died, two and a half hours later.

Burks sued the hospital, claiming both malpractice and that the hospital had violated state and federal laws requiring that borderline viable newborns receive everything the hospital had to offer. The Supreme Court of Wisconsin sided with the mother.

In cases where the neonatologist attending delivery decides that the baby is too premature to survive and refuses to resuscitate the child over the objections of the parents, courts have supported the parents. Even though the children in these cases were fundamentally too immature to survive, and even though the official policy of the American Academy of Pediatrics is that these babies should not be put through the pain of a futile resuscitation, courts have determined that parents have the right to a resuscitation attempt anyway. Repeatedly, when there is a disagreement about resuscitation, courts have sided with the advocate of life, whether that person is the parent or the physician.

Steve Ringer told me about an analogy he used to use when he counseled pregnant women who were at risk of delivering prematurely.

If he was telling a woman who was twenty-six weeks pregnant that there was a 15 percent chance that her child would have a really bad neurologic injury, he often used the analogy of the lottery.

"I would tell the parents, 'If someone offered to sell you a lottery ticket for a million-dollar jackpot and they said you have an eighty-five percent chance of winning, you would buy the ticket, right?' And of course, when I put it that way, parents would nod and smile, because who wouldn't buy that ticket?

"But then one day one of the residents pulled me aside and said, 'You know, Dr. Ringer, the problem with the analogy is that when someone buys a lottery ticket and they don't win, they go home with nothing. But with this family, if they don't win and their baby ends up with a devastating neurologic injury, they didn't just not win the lottery. They have to pay the million dollars and they will be paying for the rest of their lives.'

"I stopped using that analogy," Ringer said with a rueful smile.

It would be easy for a neonatologist to resuscitate any baby that had even the remotest chance of survival and take the position that the courts would defend the actions, and that federal law might even demand it. And to be sure, some neonatologists—mostly those guided by personal or religious beliefs—do this. Some legal scholars believe that even though the God squad signs no longer hang, the revised Baby Doe rules and the 2002 Born-Alive act demand aggressive care for all newborns except those for whom death is certain. "Under current federal law, resuscitation is mandated," said Craig Conway, a research professor at the University of Houston Health Law and Policy Institute.

But dating back to the 1984 détente, the American Academy of Pediatrics' interpretation of the law allows physicians significantly more leeway in decision making, leading to the curious situation that the field of neonatology generally functions according a "best-interests" code of ethics, one that varies from hospital to hospital and largely ignores the laws that purport to dictate how neonatologists should practice.

Because major disagreements between parents and physicians are still extraordinarily rare, no one has an interest in disrupting a system that works. Aside from the very few cases that draw legal attention, physicians and parents make appropriate decisions that are in the best interests of the babies.

At Providence St. Vincent Medical Center in Portland, Oregon, these decisions were inconsistent. "On one night we would be resuscitating a twenty-three-weeker, and on the next night we would be deciding with the parents not to intervene when a twenty-five-weeker was born," recalled Joe Kaempf, a neonatologist. "It was creating a lot of stress in our unit."

To ensure consistency in the counseling that patients received and to develop a consensus among caregivers, Kaempf led an initiative to draw up some guidelines for resuscitating newborns who were between twenty-two and twenty-seven weeks of gestational age. The process involved a series of meetings of neonatologists, obstetricians, nurses, and administrators, who reviewed the medical literature and surveyed the doctors on staff at the hospital about their practices and their opinions. Guidelines emerged based on data from St. Vincent and from nationwide studies that reflected a consensus opinion at the hospital: for babies born before twenty-three weeks gestational age, the hospital's policy was that resuscitation and NICU care was not an option, even if parents requested it.[8] For those born between twenty-three and twenty-six weeks, NICU care was optional, with the parents empowered to decide. For babies born at twenty-six weeks and later, NICU care became obligatory except in extraordinary circumstances.

It was Kaempf's strongly held belief that this hospital policy was legal—a belief shared by the hospital's legal department. "All that [the law] says is that any birth in the United States needs to be evaluated by a physician," Kaempf said. "It does not say you have to resuscitate."

Boston University bioethicist George Annas, who has written extensively on this topic, agreed with Kaempf. "The doctor is always going to have the responsibility to evaluate the kid when it's born," Annas said. "But evaluate does not mean resuscitate."

In a paper Kaempf wrote describing St. Vincent's experience with the guidelines, parents who were interviewed reported that the guidelines were clear and useful. Perhaps not surprisingly, no parents have disagreed with the guidelines, so there has been no conflict. Fully 31 percent of parents who were counseled by Kaempf and his colleagues and whose babies were delivered between twenty-two and twenty-six weeks opted for palliative comfort care.[9] Unfortunately, most hospitals don't have such well-designed policies, and decisions are made at the last minute by whoever claims to be in charge.

Janet, a good friend of mine, woke up one morning twenty-four weeks and one day into her second pregnancy with damp underwear. She called her obstetrician, who didn't take her concerns seriously. But the leaking continued, and the next day she was admitted to Brigham and Women's Hospital with ruptured membranes.

Unlike my earlier patient, Janet and her husband, John (their names have been changed to protect their privacy), understood that preterm delivery could leave them with an injured child but were prepared to accept and love the child, injured or not. Janet lay in her hospital bed receiving the intravenous antibiotics that sometimes ward off labor and the steroid injections that would mature her baby's lungs if it did not.

Ringer came to consult with them and described the significant risks that accompany a birth at twenty-four weeks. After a long discussion, Janet and John told Ringer that they did not want their son resuscitated if he was born prior to twenty-five weeks. To Janet and John, an intensive resuscitation seemed to be an unnatural intervention that went against biology. If Janet's pregnancy continued to the point where their child had a reasonable chance of a good outcome, then so be it. Before then—and they set that point at twenty-five weeks, based on counseling from Ringer—they didn't think a resuscitation was in their child's best interests.

"You know," said Ringer, "you could always wait and see how he does. We can always change direction if he is not doing well in the NICU."

"The problem," John said, "is that we can make this decision now. But once my son is born, and I see him, I would never be able to tell you to stop doing everything for him."

Visiting Janet later that day and misunderstanding Janet's and John's position, the obstetrician told her that abortion at that stage was illegal in Massachusetts but that she could travel out of state if she wanted to end the pregnancy. Janet was horrified. Like Karla Miller, Janet and John were under the impression that the doctors had agreed to respect their request that their child not be resuscitated if he was born before the twenty-five-week benchmark.

But two days later, when Janet developed a fever and went into labor, Ringer was home in bed, and the neonatologist who was on call that night told Janet and John that he wouldn't honor the plan they'd made with Ringer. *I discussed with the parents what choices they have,* the neonatologist wrote in the chart, and then he described details of the resuscitation he planned.

"We don't really have any control here, do we?" John asked him.

There was no answer.

When their son was born, that doctor intubated him and whisked him off to the NICU.

It became clear, however, that Janet's fever was caused by an infection in her uterus that had overwhelmed her son. His lungs were seriously injured, medications were required to support his blood pressure, and an ultrasound of his head showed bilateral hemorrhages. The neonatologists told his parents he was unlikely to survive.

At the last minute, the child was disconnected from the ventilator and a neonatologist pushed tiny rapid breaths into his lungs using a handheld rubber bladder while the team pushed the incubator and raced upstairs to Janet's hospital room so she Janet could hold him while he died.

"I remember," Janet said, "that by the time they arrived and handed him to me, he was already dead."

It is worth noting that once the extent of Sidney Miller's injuries was known, on more than ten separate occasions, neonatologists caring for her counseled her parents that it would be entirely reasonable to withdraw life support and allow Sidney to die. But once their daughter was born and in the NICU, Karla and Mark were attached, and each time, they insisted that everything be done to save the life of their daughter.

Ringer made his way to Larissa's station. As little as she was, she had begun to acquire the paraphernalia that accumulates around babies. She had a few outfits now in addition to the Polaroid photos taped to her incubator and her sisters' drawings taped to the wall.

With my hands extended through the portholes into the incubator, I absentmindedly held Larissa's curled-up form. One hand encircled her bottom and her feet, keeping her knees curled up under her, the other supported the length of her back and her head. Her hair looked like it might be dark red, as Kelly's had been when she was a child.

Ringer must have been gazing at Larissa for some time as he stood next to me, but with my mind elsewhere, I had not noticed him come up. So as not to startle me, he moved before he spoke. "She's doing great, don't you think?"

"Is she?" I asked.

"Well, she's off the vent." Ringer started to list the progress Larissa had already made.

"I know." I smiled tightly.

"You don't find that reassuring. Do you?"

"Not really."

Ringer tipped his head to one side, thinking. I was still holding Larissa in my hands. My eyes faced her.

Ringer began again: "I know we are trained to practice evidence-based medicine. To seek the maximum amount of objective data—labs, imaging, examination findings—and then go to the literature." He paused. "You've done a lit search?"

"Yes."

"It's not reassuring, is it?"

"No."

Ringer paused again, thinking about how to get through to me.

As physicians, we are trained to treat the parents of our patients with respect. To speak to them as equals. When the parent is a physician, it can make the conversation easier, because a language is shared. Medicine is also a bond. On the other hand, my MD was all of seven months old. There wasn't a lot of experience resting on those letters.

"You know, I've been doing this a long time," Ringer said. "And Larissa, she just looks like a good baby."

He wandered off but came back a few minutes later. "I've asked the ethics committee to help us. I think you should meet with them. Would you and Kelly be free at three tomorrow?"

The meeting took place in the NICU conference room. A white-board covered in statistics from the previous lecture to the NICU trainees about TPN for feeding premature newborns dominated one wall. Along a back table, muffins hardened next to a cold cardboard box of coffee from Au Bon Pain and some material from the Abbott pharmaceutical sales representative who had provided the food.

Kelly and I sat together near the door. On the way in we passed Ringer, who was in a heated discussion with a gaunt gray-haired man wearing a blazer. They came into the room together and sat. A social worker wandered in, greeted everyone, and started shifting papers around in her thick leather-bound organizer.

The man in the blazer, who introduced himself as an attorney and chairman of the hospital's ethics committee, started talking to me. "Dr. Ringer tells me that your daughter Larissa is doing quite well. But that her prognosis is uncertain because of her bleed."

"Yes," I said. "The literature is not reassuring." I had a paper from the library in my hand.

"I see." He looked at me. "If you didn't want to raise your child yourself, would you ever consider adoption?"

Kelly visibly gagged and then looked at me, her face white. Holding her stomach against the pain of her cesarean incision, she got up and left to vomit.

"Of course not," I said angrily, and followed her.

CHAPTER 7

Is Your Life Good?

Amyotrophic lateral sclerosis, also known as ALS or Lou Gehrig's disease, is a progressive illness in which paralysis slowly sets in while cognitive function remains intact. Patients diagnosed with the disease often recall a clumsiness—increased tripping, or worsening penmanship—that went unexplained for months. Patients with ALS eventually lose the ability to walk and become wheelchair-bound. At some point, they become unable to swallow and need a tube inserted into the stomach to receive nutrition. Speaking and breathing also become impossible, as the muscles of the diaphragm and chest refuse to function. A tracheostomy—an aperture made surgically through the neck and into the trachea—connected to a ventilator is required to prolong life when this happens. Ultimately, a locked-in state occurs in which all muscular activity, even blinking, is impossible; sight, smell, hearing, and the normally functioning mind are intact, but the individual has no way to communicate with the world around him.

The disease arrives in middle age. Patients have normal lives until that point—they have families, careers, hobbies. Unlike a child born with cerebral palsy who never learns to walk, those with ALS have the ability but lose it. Because their cognitive capacity never wanes, they can make choices for themselves. Some decide not to have the feeding tube implanted and opt to die when they can no longer eat. Most accept the feeding tube but opt not to have the tracheostomy. When the patient's breathing fails, the level of carbon dioxide in his blood rises and he goes to sleep. A morphine infusion wards off the panic that accompanies suffocation.

Time is on the side of the patient with ALS. He takes stock of his life on a daily basis and decides for himself when his quality of life has deteriorated to the point that death is preferable. A decision doesn't have to be made in an instant, or in a day.

Because the progression of ALS is slow—although inexorable—the patient can reflect on his sadness at the life he has lost while learning to appreciate the life that he still has.

Often, quality of life becomes a moving target. What was unacceptable when the diagnosis was made—a life without walking, a life without eating, a life without conversation—becomes tolerable when the alternative is death. But the decision is the individual's. The health-care proxy—the person empowered to make decisions for the patient when the patient cannot make the decisions himself—is really just an agent acting on the patient's wishes, confident that the decisions are the patient's and no one else's.

I had an image of Larissa at thirteen in a bulky motorized wheelchair padded to support the spastic limbs of a child with cerebral palsy. In my imagination, it is evening, and her sister Hannah, now sixteen, has just gone out with a group of friends. Children in Larissa's class are having a party, and Larissa has been invited, but she doesn't want to go. "Look at me," she says to Kelly and me. "Why would I go to the party?"

There are sharp differences between cerebral palsy and ALS. First, ALS is progressive—it strikes an unimpaired life and gets worse. Cerebral palsy is nonprogressive—it gets neither better nor worse over a life span that is essentially normal. Second, we had no real idea what Larissa's life would look like in five, or ten, or forty years. The neurologists' predictions were like a bad weather report: partly cloudy with a 50 percent chance of rain. The prognosis of ALS is certain—the only uncertainty is how long it will take for the disease to progress. Third, Kelly and I were making decisions for Larissa without the benefit of a discussion with her. Maybe of greatest significance, Larissa would live her life without precedent or context—she would never think back to the time before her injury or her illness.

Unfortunately it is not uncommon to watch family members or friends have accidents or illnesses, like David Carmel's, that change—diminish—their quality of life. It is an innate skill, the capacity to evaluate one's own quality of life, and a part of empathy is understanding someone else's quality of life. But it is something else entirely to try to conceptualize the future quality of life of a child injured at birth. And that was my responsibility, and Kelly's.

Recall the riddle of the sphinx: What walks on four legs in the morning, two legs at noon, and three in the evening? Millennia old, our concept of life is imbued with what is supposed to be. Dr. Elisa Abdulhayoglu recognized in herself—although she didn't yet have children at the time she took care of Larissa—the same desire that most parents have. "We all want that," she said. "You hope your child will be born weighing eight pounds, on their due date, and after you are in the hospital for a few days, you and your child go home where they breast-feed beautifully, and then they go to school and do well, and go to college, and get married, and have babies of their own."

Like the other mothers and fathers who crossed the Brigham and Women's lobby, Kelly and I dreamed about ballet lessons, homework, arguments about appropriate dress, boyfriends, and the first job after college.

Like many parents, we were willing to embrace another life, one that included disability, with braces and a wheelchair, tutors for learning troubles, surgery for muscles that contract because no nerve stimulus tells them not to.

But I was haunted by an image of my child as an adolescent locked in to an existence of Steve Ringer's making that was intolerable for her. Not for me, but for her. The neurologists had been quite clear that a life with significant disability was a real possibility, although they were unable to paint the details of that life. What tormented me was the vision of my child turning to me after Hannah had gone out with her friends and asking, "Dad, why did you do this to me?"

Once, in a restaurant over a lazy late-summer dinner, one of my parents' friends, a French intellectual with a paunch and white close-cropped hair, leaned toward my brother, who was twenty at the most, and asked, "So, Elias. How is your life?"

Nearly two decades later, this moment sticks with me because the question seemed so preposterously open-ended and absurd, because the man was so serious, and most of all because it took my brother completely by surprise. "It's good," he'd said, without thinking. "My life is good." Jean Jacques could have asked about specific criteria that make for a good life. *Are you happy?* he might have asked, or *Are you fulfilled?* or *Are you content?* or *Are your friendships rewarding?*

Although researchers do not ask, How is your life? they do ask questions about happiness and they have developed ways to measure it. Many researchers have built careers on measuring quality of life using sophisticated metrics, and many others have developed ways of asking people about the contribution of their health to their quality of life, using scoring systems to assess people's "health-related quality of life."

Groups of experts have gathered in conference centers around the world to create definitions of the word *health*. The World Health Organization, for example, defines *health* as "a state of complete physical, mental, and social well-being and not merely the absence of disease or infirmity."[1] This definition isn't much different from the one used for quality of life.

There is much disagreement about what constitutes health-related quality of life, but broadly speaking, if an illness or impairment gets in the way of a normal life, most researchers agree, then health-related quality of life is diminished.

Measuring quality of life is not like measuring blood pressure—you can't put a happiness cuff on a person's arm and inflate it. There are three ways that researchers measure health-related quality of life. My favorite is the standard gamble technique, which neonatologist Saroj Saigal at McMaster University in Ontario has used extensively to study the quality of life of children and adolescents who were born prematurely.

A hypothetical child, Sandy, is described to a parent participating in the study. Sandy has difficulty seeing, hearing, and talking. Sandy—intentionally given a name that could be male or female—walks with assistive equipment and is sometimes worried, angry, or sad. Sandy learns very slowly, needs special assistance in the classroom, and requires special equipment to accomplish daily tasks such as eating, bathing, and toileting.

The parent is then asked to imagine herself—moms are studied more often than dads, because they are more likely to bring their kids to the doctor and be available to participate in research studies—living as Sandy.

The mom is then offered this gamble. "If I told you that you could enter a lottery that would cure you (Sandy), would you do it?" the researcher asks. "You need to understand that there are two possible outcomes if you enter the lottery. There is a ninety percent chance you will be cured and live in perfect health, but there is a ten percent chance you will die." Most moms ask for the hypothetical lottery ticket.

The hypothetical gamble then changes; the chance of cure is now 10 percent and the chance of death 90 percent. In this case, most, but not all, moms choose not to play the lottery.

The researchers continue changing the odds, homing in on that balance between the chance of cure and the possibility of death until the research subject has trouble deciding whether she would play the lottery or live life as Sandy. The researchers ask hundreds of subjects about this gamble until they have enough answers to create averages and can conduct the statistical analysis that lets them trust the answers they have.

This is the utility score, and depending on whom you ask, for Sandy it is somewhere between 15 percent and 30 percent. That is, most moms would play the lottery if the cure rate was 15 to 30 percent and the risk of death 70 to 85 percent. Because of Sandy's severe disabilities, women would take a huge gamble for the relatively low chance that they would be cured.

This unemotional quantitative system allows researchers to compare different injuries and states of disability and get utility scores between 0 (death) and 1 (perfect health) that can be compared. (Some scales allow answers from -1 to 1 in order to accommodate existences considered worse than death.) The life of an amputee can be compared to the life of a person with congestive heart failure or to the life of a patient with kidney failure who requires dialysis twice each week.

The standard gamble and other, simpler, techniques to measure quality of life have been adapted for use with children by breaking down the concept of quality of life into manageable pieces. Children are asked about their ability to walk, see, hear, or go to the bathroom, and whether they are in pain, are happy, or are angry. Then the children are asked to compare their own quality of life to that of hypothetical children. Through this system of comparison to the hypothetical children, one child's perception of his quality of life can be compared to another's. When compared head-to-head, these inventory systems give results that are similar to the standard gamble method.

Dr. Saigal cuts an elegant figure; her accent is a blend of her native India, where she trained in medicine and pediatrics, and her adopted home of Ontario, where she moved to train in neonatology. She has spent years asking children who were born prematurely, as well as their parents and their doctors, about their quality of life. Earlier in her career, when she was a neonatologist caring for extremely sick premature newborns, parents often asked her something along the lines of "Dr. Saigal, if my child is very handicapped, will he say to me, 'Mom, why did you save me?' "

"I agonized over this," Saigal recalled.

In one of her studies,[2] Saigal asked the parents of a group of teenagers who'd been born weighing less than two pounds about their quality of life. Among these children, 13 percent had cerebral palsy, 10 percent had cognitive impairment, and 6 percent were blind. Most had some type of disability, although they varied in severity.

In spite of this, nearly 60 percent of the parents said their kids had "perfect health," using a 1.0 utility score on a standard gamble test. Overall, the mean utility score for these former preemies was 0.91, only modestly lower than the 0.97 given by the parents of a group of children born full-term.

When the children with neurologic impairment were considered separately, they had a lower quality of life, according to their parents, with a mean utility score of 0.78. The parents of former preemies without neurologic impairment reported utility scores indistinguishable from those reported by the parents of full-term children, even though most of these kids had other minor impairments.

Of the 149 parents of former premature newborns interviewed, only one parent—whose child had significant neurocognitive impairment—said that her child's quality of life was worse than death.

Others have studied the quality of life of children and young adults born prematurely and have seen results similar to Saigal's. These children and their parents are not viewing the world through rose-colored lenses—they see clearly that the children can do fewer things than kids born at term and recognize that they must contend with disabilities that most children do not have. When queried about the impact of their specific disabilities on their lives, these children and their parents do appreciate that life with disability is more difficult than life without such challenges.

But when asked about their overall quality of life, children born prematurely tend to see themselves as similar to those born at term, irrespective of their own limitations. Their parents tend to agree.

Saigal once asked a group of adolescents with disabilities to rate their quality of life. Then she created hypothetical children whose disabilities exactly matched the adolescents' she had surveyed, and she asked the adolescents to rate the quality of life of the hypothetical children. In almost all cases, the disabled child rated his own quality of life higher than the hypothetical child's whose disabilities matched his.

A young flutist once told Dr. Saigal, "I'm so glad I am blind and not deaf. If I were deaf, I would have died." Her colleagues report studies of disabled children who declined a hypothetical treatment that would cure them because they feared they would lose their identity if they were suddenly normal.

Of course one group of children who cannot speak for themselves about their quality of life are those children with cognitive impairments severe enough to prevent them from completing surveys. Research describing the quality of life of the cognitively impaired is scanty, perhaps because it is so difficult to do and relies on the parents' assessment of their child's happiness, health, and well-being.

Researchers have to ask their parents to speak for them, which is inherently problematic: How can an adult who relies on her cognitive capacity to get through each and every day reasonably assess the quality of life of her child whose enjoyment of the day is presumably based more on sensation and emotion—hunger, thirst, warmth, love—than on cognitive function?

Whose quality of life is better? Who is happier? The child with normal cognitive function who has cerebral palsy that prevents most voluntary movement, or the child with normal motor function whose IQ is 60? Personally, I imagine that I would rather be born with an IQ of 60 and normal motor function into a supportive family; I would find it extraordinarily frustrating to be locked inside a body that I couldn't control. But ask me which child would be more rewarding to parent, and I would take the child with cerebral palsy because of the opportunity for cognitive interaction.

My preference for a conversation over a game of catch creates a dilemma. Does it matter what parents want? When your mom has a stroke, you take care of her. When your wife develops breast cancer, you nurse her through surgery and radiation. When your child is diagnosed with leukemia, you don't consider the impact on your family when planning treatment.

I believe the answer is no—it does not matter what parents want. What does matter is that parents want what's best for their child. Of course it feels good when the coach says, "Your daughter is a great soccer player." But parenting is best done selflessly, whether the child is three months premature or seventeen years old.

This is why the attorney's suggestion that we consider adoption for Larissa was so offensive; it was an accusation that we were putting our own needs ahead of Larissa's.

No one is completely selfless. Find me that saintly parent who never considers his or her own needs, and I'll show you a spoiled child. It is reasonable to lament the burden of caring for a child with health issues—it's just not okay to make bad choices for that child, who is completely dependent on you to act in her best interests.

What about the parents? Perhaps not surprisingly, research has demonstrated that in families who have children with cerebral palsy, the integrity of the family unit is the most significant factor in determining how parents coped with a disabled child. To be sure, the child's health requirements were important, and the child's behavior even more so (what parent can't appreciate that?). But when asked, parents of children born prematurely tend to have the same quality of life as parents with kids born full-term.

There is a tendency to attribute this remarkably rosy outlook on life to resilience—that innate human ability to bounce back from adversity. Whole theories have grown up around the idea that some people are more resilient than others. How often have you heard it said with admiration about someone who has suffered some stroke of bad luck, "She's so strong"?

One physician told me the story of the mother who was devastated to learn that her child had suffered severe intracranial hemorrhages during her preterm delivery; she was counseled by physicians in the NICU that the child's prognosis was not good, that walking was unlikely, language acquisition improbable, independence unrealistic. Against advice, she demanded that everything be done for her daughter, insisting that her child was a gift from God and would live. The physician saw the child four or five years later and was saddened to see that all predictions were correct: The child rode in a wheelchair and was contorted by cerebral palsy, was fed through a tube in her stomach, and expressed pleasure or discomfort with smiles or groans. But the mother bounded up to the doctor with a huge grin and told her, "You see, everyone said how poorly she would do, but look—look at how well my daughter has done."

It turns out that resilience may not have a lot to do with it. Saigal wondered if parents would change their opinions about a disabled child's quality of life if the hypothetical impaired child became a reality. Would parents adapt to the idea of parenting an impaired child and rate that child's quality of life higher? "The perception was," Saigal said, "that parents' values change after the birth of their child."

She asked a group of pregnant women who were at high risk to deliver prematurely to rate the quality of life of hypothetical impaired children. She also asked the parents of extremely premature newborns to rate the quality of life of hypothetical impaired children. She asked this group soon after their children's delivery and then again a year later when the reality of parenting a premature newborn had sunk in, but always with reference to a hypothetical child—never their own.

Whether or not they had a premature newborn, parents were consistent in their ratings of the quality of life of Saigal's hypothetical children.[3] Their values remained the same. If resilience was a part of the equation, it was the resilience of parenthood, not the resilience of parents actually rearing an impaired child. "I really take my hat off to parents," Saigal said. "They go through tremendous trauma in the NICU, then they worry about how their child will do, and there is a long period of uncertainty. And if their child is impaired, they have to go through all of that care. But parents are so resilient. They will do anything for their children."

In another study,[4] Saigal asked adolescents, some of whom had been born prematurely, and their parents, neonatologists, and neonatal nurses to rate the quality of life of a set of hypothetical disabled children. When Saigal looked at her data, she was surprised. Her adolescents had lots of different opinions about what kind of lives the hypothetical children led, but the answers had little to do with whether or not the adolescents surveyed were disabled or born prematurely. Parents also rated the quality of life of these hypothetical children more or less the same, irrespective of whether they were the parents of children born at term or born prematurely. The median utility score for Sandy for the parents was 0.25, the same as the median score given by the adolescents.

In contrast, the physicians and nurses surveyed gave Sandy a median utility score of 0—equivalent to death. Health-care providers—the ones on the front lines fighting for the survival of the tiniest premature newborns, and the ones deciding whether or not to resuscitate these babies—uniformly rated the quality of life of Saigal's hypothetical children lower than parents did, and lower than the adolescents did. The only exception was that the health-care providers thought the quality of life for the least impaired children was higher than the adolescents did, although lower than the parents did. Saigal doesn't know why physicians are so pessimistic but wonders whether they bring their experience caring for very injured children to the experiments. "We look through different-colored lenses, I suppose," she said.

In a famous study conducted in the mid-1970s,[5] researchers interviewed three groups of Illinois residents about their happiness. The members of one group had won between fifty thousand and a million dollars in the Illinois state lottery. The second group had been victims of accidents that left them paralyzed from either the waist or the neck down. The third group was a control group of approximately the same age and economic status as the other two groups.

Not surprisingly, the lottery winners rated the experience of winning as highly positive, and the accident victims looked back on their injuries as negative events. But when asked about their current happiness, the lottery winners were about as happy as the control subjects (victims were less happy), and when asked how happy they would be in the future, all three groups believed they would be equally happy.

Interestingly, when the lottery winners were asked how they enjoyed the daily pleasures of life—hearing a joke, reading, talking with friends—they reported less happiness than the control subjects did and about the same amount as the accident victims. Good fortune seems to have soured their day-to-day enjoyment of life.

Like lottery winners and accident victims, parents of premature newborns cannot see beyond the overwhelming shock of the present to a time when the extreme situation becomes mundane, and the obvious differences routine. But this transition does inevitably occur, and whether it's because of human resilience or adaptation or just the instinct to make lemonade when handed lemons, there's a robust body of scientific literature that supports the old adage that happiness really doesn't have a lot to do with the external conditions in life.

Now, years later, Saigal feels she can answer the question that parents ask: "I can tell parents now that the children rate their quality of life fairly high, and that none have asked, 'Why did you save me?' And that is reassuring to parents."

Lying propped on her side in her incubator, her back against a towel roll, Larissa was beginning to thrive. There were no new surprises. Her oxygen demand dropped each day.

However, movement impairment seemed certain. Would she

walk? Would she use her right hand? Normal function was not going to happen. The question was how severe the impairment would be. But for Kelly and for me, the more agonizing uncertainty was the question about Larissa's cognitive function. Would she read? Would she laugh at jokes? Would she tell them?

The medical literature suggested Larissa's prognosis was grim; the best pediatric neurologists in the world had hedged their bets, and Ringer had opined that Larissa looked like a good baby but admitted that his view was based on instinct, not clinical data.

We had read dozens of published studies, talked to a host of experts, and agonized for countless hours. In the end, there was no decision. I walked into the living room one night where Kelly was attempting to distract herself by watching television. Standing in the doorway, I said, "I think we should hope for the best."

"So do I," she said.

CHAPTER 8

The Recovery Pathway

Innumerable medical school essays recount the applicant's volunteer experience with some highly empathetic medical group—the traveling troop of cataract surgeons in Guatemala, the health center in urban Baltimore, the emergency room doctors in Chicago—and identify the point when medicine became the applicant's calling. Jason Carmel's essay framed a hypothesis: that service requires both skill and compassion—alone, neither is sufficient.

The hypothesis came from Carmel's experience working at Camp Ramapo in New York's Hudson Valley. The Carmel family had been involved with the camp for decades, and Jason and his brothers had been encouraged to get involved when they were old enough to be counselors.

Ramapo took troubled children, mostly from urban environments, and created an atmosphere that provided a mainstream camp experience—but with the support and structure these kids needed. Whether the child had limited cognitive function, autism, or just severe behavioral challenges, Ramapo's staff used the tools at their disposal to help these kids—who had failed so often at so many things—succeed.

Carmel recognized that the formula for success wasn't rocket science: a highly structured environment with no downtime during which kids could get into trouble; one-to-one supervision by college students like Carmel who attended a one-week boot camp prior to the start of the summer learning Ramapo's time-perfected techniques; and nearly limitless encouragement and enthusiasm.

"I tend to be a pretty optimistic and can-do type of person," Carmel said, reflecting on the seven summers he spent at Ramapo beginning at the age of seventeen, "but having that kind of experience early on, where you see the impact you can have on another human being, is an important experience to have."

All three Carmel brothers spent summers at Ramapo. David, who shared Jason's can-do attitude, had taken the lead while he was a student at Harvard and started a Ramapo-inspired mentoring program for children identified by their Head Start teachers as being at risk for behavioral and academic difficulties. The program, called Jumpstart, originated in New Haven and Boston but grew quickly, and today it operates on eighty college campuses and serves over nine thousand kids annually.

Ramapo showed Jason Carmel what a thoughtfully designed curriculum combined with an enormous amount of compassion could do for a troubled child. He also learned what it could not do: it could not help the child who was hallucinating and needed medical care or the child with epilepsy who had difficulty learning and behaving in school because he kept seizing. To a can-do guy, these problems that could not be addressed by a structured curriculum were frustrating. They came to represent a challenge and informed the attitude toward service that he outlined in his medical-school-application essay.

Three years after taking a "year off," Carmel graduated from medical school at Columbia. With an MD and a neuroscience doctorate in hand, he had a new direction: understanding—and eventually repairing—spinal cord injury. He would do a residency in neurology, and because adult neurologists tended to focus on degenerative diseases, such as Parkinson's and Alzheimer's, and stroke, he chose pediatric neurology, which more closely aligned with his interest in repair and regeneration of damaged nerve structures.

Columbia has one of the nation's preeminent neuroscience research programs and includes among its faculty two Nobel Prize winners, but when Jason Carmel started his residency the next year, he aligned himself with Jack Martin, a lesser-known researcher who had some innovative ideas about the brain. In an era when discovering molecular and genetic mechanisms for degenerative diseases won publicity and grant dollars, Martin was figuring out how the brain and spinal cord control movement.

Experimenting with cats and rats, Jack Martin had built a career working out the structure and function of the corticospinal tract, the bunch of neurons connecting the brain and the spinal cord that is responsible for most voluntary movement of the arms and legs. At the time that Carmel came along looking for a research project, Martin was asking some fundamental questions about the spinal cord—how do animals learn how to move? how do they relearn how to move after injury?—and he was developing some intriguing hypotheses. For Carmel, a newly minted neuroscientist with a twin brother in a wheelchair, these questions held both intellectual and personal urgency.

The cell bodies of corticospinal tract neurons are located under the surface of the cerebral cortex and serve a variety of functions, including movement planning, integration of sensation and action, and perhaps even processing of emotion and cognition. At least half of these cells are located deep inside the precentral gyrus of the cortex, approximately three inches above the ear. These are the neurons that constitute the homunculus, the odd map of the human body superimposed on the cortex that shows that more of the brain is devoted to highly innervated areas, such as the hand, than to less dexterous anatomy, such as the elbow.

The long axons of these cells coalesce downward toward the brain stem and serve as the principal pathway for voluntary movement. At the brain stem, most cross over to the opposite side, which explains why the left side of the brain controls the right side of the body and vice versa, although a small number of cells send their projections down the same side of the brain stem. After they cross, the corticospinal tract axons descend down the brain stem in bundles reminiscent of pyramids, giving rise to their alternate name, the pyramidal tract. These cells continue down the spinal column and then connect at synapses with one or more lower motor neurons, which branch out into the body to control various muscles—for example, a portion of the muscle involved in rotating the palm of the hand toward the sky.

Anatomically, the fact that some, but not all, axons—which are each as thick as a human hair and several feet long—cross from one side to the other is critically important. It explains why an injury to the left side of the brain, as happened in Larissa's case, affects the right side of the body.

The small set of nerve fibers that never cross influences the unpredictability of brain injury as well as the brain's capacity for recovery, because this group of nerve fibers represents a direct connection between the right brain and the right side of the body.

Early in their development, the corticospinal axons branch wildly at the level of the spinal cord where they end. None of the connections made by these branches is particularly strong or robust, and over time these branches are pruned back, leaving a few primary connections that grow in strength.

What causes this to happen? Is it hardwired—some complex set of genetically programmed instructions relayed to the neurons by growth factors and neurotransmitters? Or is it plastic—that is, does it depend on the experience and external environment of the child?

Martin had a theory; he thought that at birth each side of the brain was more or less able to control either side of the body—or both sides at once. He believed that the pruning of branches over time represented competition, one group of neurons beating another to set up productive connections. During the process of normal development, the corticospinal tract fibers that cross from the right brain to the left side of the spinal cord are more robust than the thin tract of neurons descending on the left side, so the right-side fibers outcompete the same-side fibers in making meaningful connections to the left hand and left leg.

Evidence for his theory is in every newborn nursery: babies' early movements are frequently mirrored—that is, the right side mirrors the left—and it is only over time that purposeful, single-side movements develop and are refined.

In a series of experiments, Martin and his students set about proving his theory. They kept asking, "What is the exact pathway here?" Then they would arrange experiments to find the answers.

Martin was also investigating connections, but his work focused not on the cortex, where the signal originated, or the insulating white-matter cells, whose failure was responsible for cerebral palsy, but rather on the wires of the corticospinal tract that carried signals from the cortex down the spinal cord.

In one experiment using a cat, the researchers sent a mild electric current through the corticospinal axons as they passed through the cat's brain stem and then measured the effect on the muscle innervated by that axon. They found that once muscle stimulation had occurred, it took less stimulation the next time to achieve the same effect. They used the neurology term *facilitation* to describe this process that really represented learning on a cellular level.

Recall the example of learning to type. When I was learning to type, I formed, for instance, the image of an *X* in my mind. Then I looked at the keyboard and saw that the *X* required that I move the ring finger of my left hand down to the row below the home row, where my finger was poised above the *S*. So I moved that finger to strike the *X*. It was a slow process, requiring multiple steps and involving numerous errors. Over time I learned fluency, and the *X* became second nature, so that now as soon as I think about typing *X*, my finger goes there. This is oversimplified, because centers of language, memory, sensation, and motor activity all contribute to this process, as do a complex set of neurotransmitters and receptors. But part of that learning process was the strengthening of a connection between a corticospinal neuron and the lower motor neuron that linked to a muscle that flexed my ring finger. Over time the action became easier, and on a cellular level, it required less stimulation to achieve. The behavior had been learned.

In another set of experiments quite similar to the Hubel and Wiesel research on kittens' vision, Martin's group temporarily knocked out function of the corticospinal tract on the right side of a rat brain by infusing muscimol, a medication that blocked all those nerve fibers' activity, for two weeks during the critical time of development when refinement of the corticospinal tract normally takes place. (The human equivalent to this would be paralyzing one side of a child's brain between the ages of four and fourteen and then allowing the brain to resume functioning.)

Not surprisingly, the connections observed in the rat's spinal cord on the left side, the side affected by the paralyzing infusion on the right, withered or failed to develop with the exuberance of the corticospinal connections originating in the unscathed left brain. A period of learning had been disrupted.

Martin used a piece of software called Spaceballs, named after the Mel Brooks movie parody, that created a three-dimensional matrix of virtual spheres and then counted the number of axons that passed through each sphere.

With the Spaceballs results in hand and screen projections of the fine, butterfly-shaped slices of spinal cord that showed the bright lines and dots of the fluorescent-dye-marked connections, the researchers noticed something unexpected: corticospinal connections from the unscathed, unblocked left side of the brain were particularly prominent on the left side of the spinal cord.

Usually, these same-side neurons are relatively inactive, since they're competitively overwhelmed by neurons from the larger and more prominent tract that crosses from the other side. But because the crossing neurons had been paralyzed during a critical period of development, the same-side neurons had blossomed and formed a robust set of connections in the atypical environment where they could compete and win.

Martin's papers were not going unnoticed by scientists who studied cerebral palsy. Researchers working with children who had CP had found that gentle stimulation of the injured side of the brain using a noninvasive technology called transcranial magnetic stimulation (TMS) elicited minimal response on the opposite side, owing to the injured brain tissue's inability to communicate. However, in these same patients, TMS to the uninjured side caused both sides to respond—evidence that the uninjured side of the brain had developed the normal connections to the opposite side in addition to the unusual same-side connections.

Other experiments showed that learning didn't happen simply from the brain down; the experience of the limbs mattered too. When researchers injected botulinum toxin (Botox) into the paw of a kitten at the same critical point in development that Martin had paralyzed the rat brain, they saw the same absence of normal connections. Learning depended not only on the repetition of action but also on the receipt of feedback in the form of sensations each time the action was executed successfully. (Of course this is obvious: no one improves his tennis stroke merely by going over it in his mind. But science depends on breaking down results into mechanisms, so these experiments formed the foundation for the critical research that followed.)

All of these experiments informed Martin's hypothesis: neurons compete with one another to form connections, a competition that rewards activity and punishes both injury and paralysis. A sort of Darwinism of connections is at work, resulting in profound consequences for therapy.

When an enthusiastic young resident with a doctorate in neuroscience walked into Martin's lab looking for a project, Martin and his team were wondering how they could translate their understanding of corticospinal tract development into therapies to treat patients with cerebral palsy and other motor abnormalities.

CHAPTER 9

Making It Routine

I was on a hastily arranged paternity leave from my residency. Kelly had been on the interview trail for internships in clinical psychology the week before, and now she canceled her remaining interviews. Ten days after Larissa was born, I took Grace and Hannah to school and came home to a house that was quiet and eerily calm. Kelly got ready—the cesarean incision was healing but it made her move slowly—and we drove to Brigham and Women's.

The day was bright and cold. I dropped Kelly at the front door to the hospital and she waited for me inside while I parked in the residents' lot. Although I'm certain it was the same, the lobby seemed quieter that morning except for the calm buzz of purposeful activity.

We sat in the NICU waiting area while the secretaries checked to make sure the medical team was done rounding. Parents weren't allowed in to the NICU until the physicians had checked in on each baby and made a care plan for the day.

We were buzzed through the double doors into the unit. Walking to Larissa's bedside, I thought the NICU seemed calmer too.

"Well, hello." We were greeted by Gabi Harrison, an experienced NICU nurse who had been assigned as one of Larissa's primary nurses.

"Larissa had a really good night," she said. "Her weight is finally up a little bit, and as you can see, she is doing really well with her nasal cannula." Larissa now received only a little bit of oxygen into her nose but was doing all the work of breathing on her own.

"Would you like to hold her?" Gabi asked.

Kelly and I looked at each other, caught off guard. Was it safe? Would we hurt her? We were used to the idea of Larissa in her incubator; the idea of holding our daughter was frightening.

Gabi helped Kelly position herself in a reclining chair. Then she expertly wrapped Larissa in a white hospital-issue blanket, making sure her head was covered in a cap; bunched together the intravenous tubing and cords attached to the monitors displaying her oxygen level, heart rate, and breathing; carefully lifted her out of the incubator; and gave her to Kelly.

Larissa lay in the crook of Kelly's arm. At first, Kelly barely moved, clearly afraid she would harm the tiny child. Slowly, when nothing untoward happened, she relaxed. A worried look turned to a smile. She stared into Larissa's face as Larissa opened and closed her eyes. Occasionally, Larissa would twitch against the blankets.

Then the alarm sounded, a high-pitched beeping noise, accompanied by a blinking red light. Gabi, Kelly, and I looked at the monitor. Larissa's heart rate had dropped into the eighties—way too slow for a premature newborn. A look of panic crossed Kelly's face. She looked at me. I looked at Gabi.

Gabi's face wasn't smiling, although she didn't seem worried. But the alarm kept ringing, and finally Gabi started moving toward Kelly. Then, as suddenly as it had started, the alarm stopped. The heart rate was now 124.

When Kelly went to the bathroom, I took her spot and held my daughter on my arm against my chest, her head at my elbow. Her eyes closed, and she seemed to sleep, though she still had the occasional shivering twitches. I remembered feeling similar movements through Kelly's abdomen only a couple of weeks earlier.

Toward lunchtime, we left. We ate at a diner near our house. We were out of the cold, and the pastrami Reubens tasted really good; around us, people on their lunch breaks talked business and complained about their bosses. When we finished, I dropped Kelly at home to rest and went to pick up Hannah and Grace at school.

The next day we did the same thing. We were developing a routine. It wasn't the one we had imagined, but we were becoming comfortable in it.

Right after he took the helm of the Brigham NICU, Steve Ringer made a major and unorthodox move to improve the quality of care in the unit: he announced that he would be a full-time Brigham and Women's employee and would spend all of his clinical time at that hospital only. Then he hired a deputy, who would also spend his time exclusively at Brigham and Women's.

Over the following year, he made the staff physicians choose: they could continue to rotate through other NICUs—in which case they would lose their Brigham and Women's privileges—or they could work exclusively at Brigham and Women's.

One physician, significantly senior to Ringer, challenged him. "You're not going to tell me what to do. This is the Joint Program in Neonatology—you'll have to answer to them." The physician stormed out of the Brigham NICU and walked across the street into Children's Hospital and then into the office of Michael Epstein, director of the joint program and also, temporarily, Ringer's boss.

"Ringer's destroying the joint program," the physician said. "Medicine at Harvard only works when we work as a team."

"Ringer is right," Epstein said. "You need to choose."

"No, you need to choose. Ringer or me."

"I already chose Ringer."

Ringer's team began to develop protocols for handling multiple clinical issues, even those as straightforward as ventilator settings. In the basement conference room, Ringer would corral as many neonatologists as he could—there were always some who were away working on research or at home after a busy call night. The group would review a recently published article on the topic at hand—for example, a paper that suggested a novel way of managing ventilator settings—and then a lively discussion would ensue.

Invariably, at some point in the discussion, one of the more senior physicians would get fed up.

"I understand the data we are discussing today," a veteran physician typically began, "but I'm not certain that it is appropriate to generalize these results to all of our patients."

"Can you give me an example?" Ringer asked, knowing where the discussion was headed.

"Remember that Redmont baby?" There were a few nods around the table. "That child was just not thriving until we upped the oxygen."

"Is there a published protocol we could review?" Ringer pressed.

The physician pointed to his head. "Steve, I've been doing this for twenty years. Do you want to tell me experience doesn't count for anything?"

Sometimes Ringer pushed the point. Other times he didn't. But the message was always the same: care would be based on medical evidence, and there would be a Brigham way, not twelve different ways depending on which doctor was on call.

"I have twelve neonatologists on my staff," Ringer complained, "and each one is unequivocally the single best neonatologist in the entire world."

Within a year, an order developed. "We all got to know each other, and although we didn't practice exactly alike, we practiced similarly. I could speak to a family and be confident that my advice would align with the advice my colleague had provided earlier in the day," Ringer said.

Complications seemed to decline. There were fewer pneumothoraces—injuries that occur when the ventilator punches a hole in the lung and air escapes into the chest cavity—and fewer infections.

Nursing turnover declined as well, and the staff thrived in the setting of consistent medical protocols and even-keeled leadership. Nurses could focus on their tiny charges instead of spending time and angst interpreting changing instructions from a large and transient group of physicians.

Not long after Ringer took charge of the NICU, the ophthalmologist who treated most of the smallest preemies came to see him. The ophthalmologist had just given a talk at the Perkins School for the Blind, in nearby Watertown, Massachusetts.

Using a PowerPoint presentation, he'd walked the parents through slides that demonstrated a tragic, albeit well-intentioned, mistake in the history of newborn medicine: efforts to maximize the oxygen levels of premature newborns had unwittingly caused blindness in a large number of children. When research revealed this association, neonatologists had ratcheted down the oxygen levels, and blindness became quite rare.

"Now we see very low levels of severe retinopathy of prematurity," he said, describing the condition that had historically led to blindness. "For the past few years, we have been on the cutting edge of this science." He showed recent data, which demonstrated a marked reduction in blindness, and explained that although some of the smaller preemies did get retinopathy of prematurity, it almost always resolved without long-term damage to the children's vision. He praised Ringer's new initiative and spoke about the new cohesion of his team and the protocols for use of the ventilator in the Brigham and Women's NICU, which had specific targets for oxygen levels.

A hand went up in the back of the room. "Doctor, thank you for this presentation. I'm sure it's coincidental, but you took care of my son, and he had stage-four ROP," a woman said, using the acronym for retinopathy of prematurity, "and it didn't resolve."

Another hand went up, and then another. There were five parents in the room whose children had been born prematurely at Brigham and Women's and were now vision-impaired because their ROP had not resolved. The ophthalmologist was stunned.

"Are you seeing a lot of ROP in your follow-up clinic?" he asked Ringer back at the hospital.

"Uh, I don't actually see these kids in follow-up," Ringer admitted. "Aren't you seeing the kids with ROP after they go home?"

"Only if they make an appointment."

Ringer realized there were challenges much bigger than those on his unit. These parents were describing children who were counted among the seeing when they left the NICU, but because no system was in place to keep track of the babies, the long-term consequences of the care the babies received early in life was lost to the people who most needed to know it.

"You know what these little guys really like?" Gabi said to Kelly. "They love to be skin to skin with their moms. It really calms them down."

Kelly was wearing a sweatshirt, so Gabi brought her a hospital gown and helped Kelly change into the gown so that it opened in the front. Kelly settled into the recliner, and Gabi carefully lifted Larissa out of the incubator and put her on Kelly's chest.

She expertly removed Larissa's onesie, leaving her in her tiny diaper, and curled her, chest to chest, on Kelly. Then she covered Larissa with a blanket. The tubes and wires fed out of the bottom of the blanket.

Kelly tried not to move, but Gabi showed her how to cup her hand to support Larissa's bottom; Larissa fidgeted, but then, as if she knew what she was supposed to do, she settled down.

This time, her heart rate stayed in the normal range, and her breathing continued steadily. Kelly held her daughter, stealing glances at the monitor every few seconds for confirmation that Larissa was fine.

In the early 1970s, the newborn ward at the San Juan de Dios Hospital in Bogotá, Columbia, filled all of its incubators and had nothing to offer premature newborns, who needed to be kept warm.

Families and physicians had no choice: mothers, fathers, and even extended family members were instructed that their fragile children were to be laid upon their bare chests and covered with clothing or blankets to keep them warm until incubators became available.

Dr. Edgar Rey, a pediatrician at the sprawling hospital, noticed something intriguing: babies who'd spent their first weeks on their mother's chests were doing better than babies raised in incubators.

Rey called it kangaroo care, after the marsupial practice of carrying the joey in a warm and protective pouch after birth, and the name stuck.

It is intuitive that babies who receive kangaroo care thrive, given their proximity to the sound, smell, and warmth of their mothers. The neural pathways of children deprived of attention early in life are aberrant, so it makes sense that newborns stuck in sterile plastic boxes don't do as well as those who spend their first weeks on their mother's chests.

Medical statisticians have since provided numbers that back Dr. Rey's keen observation. Babies who receive kangaroo care have fewer infections and go home sooner than babies cared for in the usual way.[1] There is even some evidence that they gain weight more quickly and might be smarter at one year of age than other babies.[2]

Of equal importance, kangarooing was something that Kelly could do for Larissa.

After I went back to work, on those cold mornings in February and March, Kelly would drop Grace and Hannah at school and then head across town to the NICU.

A nearly religious coffee consumer, Kelly now drank nothing in the morning, knowing that liquid would make her have to pee, which would mean disturbing Larissa.

Each morning in the NICU, Kelly would hang up her winter coat, carefully wash her hands, and go to Larissa's space. In anticipation of her arrival, the nurses usually had pulled the NICU's version of a Barcalounger up near Larissa's incubator. Wearing a sweater that buttoned in the front, Kelly, with help from Larissa's nurse, would slide Larissa between her breasts so Larissa's head rested on Kelly's sternum.

"It's you and me," Kelly would mutter as the day crept past.

It was all she could do. Larissa was still fed by feeding tube, so it would be some time yet before Kelly could give her a breast or a bottle.

At lunchtime, Kelly would eat a sandwich she had packed and go to the bathroom. Larissa's diaper was changed, and then they would settle in again for the afternoon. When the winter light dimmed, Kelly and Larissa's nurse would put Larissa back in the incubator for the night, and Kelly would head out to retrieve Hannah and Grace.

Although we didn't do it with Kelly's consistency, the rest of the family got into the act. Between shifts at the hospital, I would kangaroo Larissa, and Kelly would later joke that she knew when I had been there because Larissa would return to the incubator with my chest hair stuck in her little hands. Even my parents took turns.

At one point, a disagreement developed. The nurses were doing the usual three o'clock shift change one day, and they sat in a group near Kelly, who was kangarooing Larissa. Babies were being discussed, and confidential medical information was being shared. One of the nurses told Gabi she didn't feel comfortable doing the sign-out—the exchange of medical information—with Kelly sitting there. She would have to leave until sign-out was over.

"It's policy," that nurse said. "Parents can't be here for sign-out. It's inappropriate for them to hear about the other babies. And besides, how can we be candid if the child's parent is sitting there?"

"We're not going to interrupt Larissa just so you aren't inconvenienced," Gabi said.

"But it's NICU policy," the other insisted.

"Let's talk to Steve." Angrily, they sought out Ringer, who was chatting in another part of the unit.

When the two sides of the debate were laid out, Ringer thought about it.

"If we encourage kangarooing, maybe we won't have to replace those incubators," Ringer joked, to defuse the tension. Then, taking a more serious tone: "It's good for the babies, and all of us want to be more family-friendly, right?"

There was general agreement.

"Maybe we could do sign-out across the room," Gabi suggested.

"I suppose we could still watch the monitors from there," the other nurse conceded.

As the weeks went by, Kelly began to blend into the fabric of the NICU. She learned about the nurses' children and met the parents of other babies. At times, nestled into the Barcalounger and covered with blankets to keep Larissa warm, she seemed to disappear to the NICU around her.

One morning there was a new arrival: a little boy born at thirty weeks named Deshane. Later that day Deshane's mother and father came to visit, carrying with them the complete bewilderment of new parents who were unprepared to be parents, let alone parents of a sick child in the NICU.

The father, a thin man no more than nineteen, had an attitude, although it seemed he was really just scared for his son. "Why does he need all those tubes?" he asked. "Why does he have those blue lights but none of the other babies have them?" The baby's nurse explained that the lights helped prevent a dangerous buildup of bilirubin, a substance that caused brain damage if it became too concentrated. "Are you saying he's going to be brain-damaged?" the young man asked.

The mother, Ciarra, who Kelly later learned was seventeen, kept looking at her son and said nothing. She didn't touch him, and when the nurse showed her how to open the incubator porthole and put her hand inside, she recoiled.

The next day, Ciarra held her son, stiffly, and for only a few minutes. The father stood back, looking sullen, and didn't say anything.

Then they were gone.

Two days later, Ciarra was back, wearing loose-fitting jeans. Now alone.

From across the room, the nurses stared. Deshane's nurse touched Ciarra's arm in a kind gesture. The two of them conversed in low tones, and then the nurse helped Ciarra position herself in a chair and lifted Deshane out of the incubator so she could hold him. Tentatively, Ciarra held her son, smiling occasionally.

Soon Ciarra developed her own routine. Every afternoon she arrived after school, lugging a heavy backpack. She'd take out a sheaf of paper or a small book and lay it on the side table. Then she would pick up her son and make them both comfortable in the chair.

"How was school?" her nurse would ask.

"All right," she would say.

One by one, or in pairs, the nurses would come to visit. They brought baby hats and clothing left by previous NICU graduates that they had taken home and washed. Signs and photographs adorned the incubator and the wall behind it.

"He's adorable."

"What a charming young man."

"You're doing such a great job, Ciarra."

Before Ciarra left each afternoon, she would wave across the room to Kelly, a small, shy wave.

Larissa was growing now, gaining approximately half a pound each week. Gabi taught us how to bathe her, first in a tiny basin and then in a larger container when she outgrew that. One morning Kelly came in and Larissa was drinking from a bottle. When Larissa was able to take all of her formula that way, the tube that snaked into her stomach was taken out and, except for the wires that monitored her heart rate and breathing, she finally looked normal.

The feeding process was involved—the eating was slow, and then Larissa had to be held upright for an hour to reduce the chance that she would vomit up the formula—but it was one more thing that Kelly could do for her daughter during the hours she spent in the NICU each day.

"They look fabulous. When can we take them home?" Fred Bobo said. His wife had delivered twins at twenty-nine weeks, and they were now all of four days old.

"Oh, Fred," Mary said. "They look just like you." She bent in to open an incubator.

Their nurse came running over. "Hold on a second, dear. I know you want to hold them, but they aren't really ready for that yet." One of the twins was still receiving breathing support.

From her chair fifteen feet away, Kelly watched the enthusiastic parents.

"Can we take that tubing off for a photo?" Fred asked, gesturing toward the breathing apparatus.

"Not just now," the twins' nurse said patiently.

A small throng of nurses gathered across the room and laughed at their colleague's situation.

For a few days, the parents didn't come into the NICU. But the next Saturday, they showed up with at least six others—two older siblings, grandparents, and a couple of friends—whom they had brought to meet their babies. The twins' older brother, who was about five, had a red nose that was running badly. The older sister, about three, dragged a dirty blanket behind her. The grandfather reeked of cigarettes.

They crowded around the incubators, popped open the portholes, and stuck their hands inside to touch the twins. The sister started to push her blanket through the porthole.

"That is so sweet, Martha," Mary Bobo said. "Are you loaning your new sister your blanket?"

"Maybe they should wash that filthy thing first," Larissa's nurse muttered under her breath.

Mary noticed Kelly and came over to chat. "How old is your baby?"

"She's seven weeks old," Kelly said politely.

"Does she have any problems?"

Larissa's nurse came to the rescue. "Ma'am, I'm sorry, but I have to ask you to respect this mother's privacy."

"Okay," Mary responded good-naturedly. "I was just being friendly." She returned to her family.

The group finished admiring the twins and got ready to leave.

"Now give your new brother and sister big kisses," Fred told his older children.

The little boy with the runny nose reached in for his brother's hand and gave it a slobbery kiss. Then he wiped his nose with his hand. Across the room the scrum of nurses cringed.

The family thanked the nurses loudly and enthusiastically for taking such good care of their children.

"Those kids are going to need a lot of luck," one of the nurses said when the mob was gone, "with parents like those."

"Oh, Louise, they're sweet," another said.

"They are," Louise said, "but they are awfully stupid."

"It's true. I'm going to pray for those babies once they go home."

April arrived, and Larissa's weight had doubled; she was now nearly five pounds. There was talk of sending her home.

But there were several hurdles—determined by Ringer and his group—she had to overcome. She spent an hour in a car seat on the NICU floor with an oxygen sensor attached to her foot to make sure that she could safely ride in a car and maintain her oxygen saturation in a reclined position. Even with the harness straps as tight as they could be, the car seat still seemed big for her, but the NICU staff assured us it was safe.

Appointments were made with Larissa's pediatrician and with the special follow-up clinic for premature newborns who had spent their first weeks at one of the three Joint Program in Neonatology NICUs. Although each was autonomous, the units still collaborated closely on numerous topics. We also made an appointment for her to see the ophthalmologist in six months so he could check on Larissa's very mild retinopathy of prematurity. She would not fall through the cracks like the children Ringer had heard about at the Perkins School for the Blind.

Then, on the eve of her discharge, Larissa had a spell: she momentarily stopped breathing. The discharge was canceled, and Kelly and I left the NICU in tears.

But a week later, we packed up the basin, the photos, the cards, and the small plastic chest of drawers we'd gotten for her clothes, put Larissa in a warm suit, and placed her in her car seat.

Seventy-one days after her precipitous delivery, Kelly carried Larissa out of the NICU and down the elevator while I pushed a cart with all of her stuff on it.

Inside the main entrance to the hospital, I left Kelly with Larissa and her stuff and went to get the car. Kelly was bundled in a winter coat to ward off the unseasonably cool spring weather.

While she was waiting, an orderly passed her pushing a woman in a wheelchair who was smiling at the two-day-old baby in her lap. Her husband pushed a cart like mine, but his was laden down with flowers and cards of congratulations.

The woman looked up at my wife and daughter and recognized there was something incongruous about them and their paraphernalia, so different from her own. She smiled but didn't say anything. Kelly forced a tight smile, and then looked away.

CHAPTER 10

Gains of Function

Many times each day during Larissa's first summer, Kelly would position Larissa on her back on a blanket spread across the living room rug. She would smile down at Larissa and then shake a brightly colored rattle just out of reach off to one side. Naturally, Larissa would turn her head and smile at the rattle and reach her little hand toward it.

"Look at that, Larissa," Kelly would coo. "Good girl."

Larissa was equally interested in the rattle whether it was shaken to her left or to her right, and she would quickly turn her head either way.

It was clear, however, that her reach was more purposeful, more coordinated, when the rattle was on her left. This was a subtle early sign of a right-side disability. What was unclear was how significant this disability would be.

Undeterred, Kelly would push Larissa's bottom to tip her so her whole body faced the rattle. Then she would bend the top leg at the knee so the knee faced the toy, bringing the weight of that leg across toward the rattle. Invariably, Larissa's top hand now reached toward the rattle, and she would tip over onto her stomach. Kelly would give that hand the rattle. "Great job, Larissa," she would congratulate her.

At first the sudden shift onto her stomach would startle Larissa, but the negative sensation of tipping was tempered by the success of having the rattle in her hand. A few minutes later, they would do it again.

Soon, Larissa was throwing her hand over her body toward the rattle. But that alone was insufficient to get the toy. Next, she would bend her knee, but since she was lying on her back, that motion served no purpose. Kelly would gently encourage the motions, always offering support and putting the rattle in Larissa's hand at the end.

One day, five months after she came home from Brigham and Women's, Larissa simultaneously turned her head toward the bright shaking rattle, bent her knee, and threw her arm. Suddenly she was on her side. She lunged with her hand, now closer to the rattle, tipped onto her front, and got the rattle. She shook it and then stuck it in her mouth.

Kelly cheered, and tears welled up. Larissa had rolled over.

It was August, which meant that Larissa was nearly eight months old. However, until premature babies reach the age of two, pediatricians evaluate their development in terms of corrected age, meaning the age the baby would have been if he or she had been full-term. Though she was almost eight months old, her corrected age was five months, which meant that Larissa had rolled over on schedule. But to accomplish that task she'd required the kind of coaching and methodical practice that most parents of full-term babies didn't have to think about. With the help of Susan Lynch, a gifted Early Intervention specialist whom Kelly hired to help tag-team Larissa's early development, Kelly plotted how to make sure Larissa developed the skills most parents take for granted in their children.

Leveraging the exuberant neuroplasticity of childhood is a very mainstream concept. Online merchants now sell special speaker systems that expectant parents can use to play Mozart to their babies in utero, the theory being that exposure to the music can influence the development of neurons even before birth. A successful maker of toys for babies and children calls itself Baby Einstein; the implication is that use of the toys will make babies smarter.

One of the most fascinating recent trends is the practice of teaching babies to sign. The idea is that babies are capable of communicating before the muscles that produce speech can generate words. By signing, babies are empowered to communicate sooner, engage with the world around them sooner, and maybe become smarter. (One dad I know observed that by unleashing communication, his daughter was also able to berate him sooner, signing, *Juice, juice, juice, juice, juice* until the cup was produced.)

These products and services are big business—there is even a company that bills itself as the Baby Sign Language University—but there isn't a lot of evidence that kids exposed to these products and services become brighter, more capable children or adults.

I remember Hannah learning to walk, scooting around the dining room table, holding on to chairs. I remember those first tentative steps and her consternation when she fell down on her bottom. I know she rolled over, sat up, and met all of her developmental milestones in approximately the right order and at approximately the right times. But I have no memory at all of the first time she rolled over, let alone how she did it. Most parents remember their children's first steps and probably know these are supposed to take place around the time of the first birthday. But unless the pediatrician turns and says, "You know, little Johnny isn't developing as expected," parents love and praise their kids and play with them, and development just happens.

When Kelly and I gazed down at Larissa lying on a blanket or asleep in her crib, we knew there would be no "development just happens" part. The neurologists had made a specific prediction: there was certain to be movement impairment on her right side. There was also a 50 percent chance of cognitive impairment. What this impairment would look like was entirely uncertain. Would her right hand be affected? Her right leg? Both? Would she have learning disabilities? Autism? Language impairment, or severe intellectual deficiencies? We could not reasonably expect her to be normal; the only question was what the impairment would be.

The answer to these questions depended entirely on neuroplasticity. Larissa had a brain injury, and so the work of the destroyed neurons and insulating cells would be taken up by other areas of her brain or it would not get done.

Neuroplasticity is more than a concept for children who have cerebral palsy—a host of studies demonstrates that healthy brain cells take up the work of damaged brain cells. There isn't any published evidence showing that motivated parents can coax extra neuroplasticity out of their child's uninjured brain cells, but what other option did we have? If there was a window of neuroplasticity—or at least a window of optimal neuroplasticity—then we had to race to take advantage of it before it closed.

We looked at the littlest things: Which thumb did Larissa suck more? When she first lay on her back and waved her arms around, did she wave one more than the other? (It was a silly question, since many baby movements are reflexes that have no input from the cerebral cortex and so don't reflect injury to those cells.)

And we wondered about her cognitive capacity. When she first started looking at us, was that a good sign? When she smiled, was that an indication of intelligence? Most of all, we waited for speech.

The newness of being home wore off, and a routine set in. I was back at work at the hospital. Grace and Hannah were back at school. Kelly was at home, often alone, with this special baby who slept, cried, and pooped like other babies. Like all moms of babies, Kelly was exhausted. Compounding the problem was that Larissa had to be fed, burped, and then held upright for an hour so she wouldn't spit up her meal. Feeding was slow, and in the middle of the night, the hour Kelly spent sitting with Larissa upright on her chest was interminable. Then Larissa might sleep for just a little while before it was time to begin the process all over again. Kelly felt like she lived in dreamlike state of sleep-deprived delirium.

Not long after Larissa came home from the NICU, I had a week's vacation, and I took a turn doing the night duty. Kelly slept through the night for the first time in weeks. The next morning she came downstairs and looked around, smiling.

"I feel completely different," she said.

"How so?"

"It's like I can think clearly. And see clearly. My head is clear, and I don't have a low-grade headache. I think I was so tired I had lost perspective on what it's like to feel normal."

———

One night in May I came into the house ecstatic about a surgical procedure I had learned that day.

"I did my first hysterectomy today," I announced proudly.

"That's great," Kelly said, without a lot of enthusiasm.

"One of the attendings typically takes the intern through the case, and today he asked me to scrub in. It was amazing."

"Hannah and Grace are asleep, but do you want to see Larissa? I'm going to take a bath. I haven't had a chance today. Now that you're home, I think I'll go do that, if that's okay with you."

"Is everything okay?"

"No, it's not. You get to go and do your hysterectomy and talk to people and chat and laugh, and your career goes on. You're focused on your hysterectomy, while I'm focused on our daughter's development. I'm stuck here in the house with no one to talk to because Larissa can't go outside yet, and I'm alone in my concern because you've moved on. No, it's not okay."

"I see what you mean."

"No, you don't. You don't see what I mean." She stormed upstairs.

I picked up Larissa, who was smiling in her bouncy seat. I decided to tell her about the hysterectomy. "So we took the round ligaments, and it was amazing—as we dissected into the broad ligament there was practically no blood. Then we took the infundibulopelvic ligaments, clamped them, and tied. Again, no blood."

Kelly returned briefly. "It's great that you are excited about your surgery. But I spend day and night with our child. I play with her. I sing to her. I talk to her. And I'm constantly wondering—all the time—what's going to happen, and what I can do to make it right. And you get to think about your surgery. It's not fair." She went off to take a bath.

The next week, Larissa went to see her pediatrician, who was pleased to report that Larissa had gained weight and was generally doing well. She told Kelly that she thought it would be okay if Larissa left the house, although she shouldn't be around other children, as they were likely to spread infections.

My cell phone rang. "Larissa can go out," Kelly said. "I made a reservation at Legal Sea Foods. We're going out to dinner. And I'm going to have a glass of wine."

We met at the restaurant. Larissa was dressed in one of the outfits we had been given as a baby gift. Grace and Hannah had Shirley Temples; Hannah's came in a plastic cup with a lid. Kelly had a glass of Shiraz. Around us people paid little attention to our family except to make occasional comments about how cute our baby was.

"She's adorable," an older woman said. "She's awfully small to be out and about. How old is she?" she asked about our nearly six-month-old baby.

"She's two months old," I lied, lacking the energy to explain the truth to this kind stranger.

"Well, she's very sweet. Congratulations."

"Thank you," I said.

We didn't wait for Larissa to reach milestones; we actively pushed her toward them, like cheerleaders and coaches for her functioning neurons. We were always looking critically at where we were and constantly scanning for delay or failures. Kelly, Susan, and I (to be fair, mostly Kelly and Susan) read articles by physical therapists who worked with disabled children, and we learned how to break down complex motions into their component parts. Then we taught Larissa, one step at a time.

When she could sit, we worked on crawling. At first, Kelly spread out a sheet on the living room carpet and placed a prized toy a few feet away. Larissa's response was to grab the sheet and pull it toward her, moving the toy to within reach. Kelly and Susan laughed at her ingenuity, but they also noticed that Larissa was using her left hand to do most of the pulling; her right hand merely assisted. Nevertheless, they exchanged the sheet for a mat that Larissa couldn't move.

Breaking down crawling into its component parts, we supported Larissa while she balanced on her hands and knees, rocking back and forth, and then helped her to move first her right hand, then her right knee, next her left hand, and finally her left knee.

"Go, Larissa."

"You can do it, Larissa." We would shake the rattle from across the mat, and Larissa would look up and croon.

There was trial and error, of course, but this was directed trial and error. By about nine months of age, Larissa was crawling.

The months flew by, and Kelly's and Larissa's routine came to include visits with an expanding list of physicians and therapists to help with fine and gross motor skills. Because of her prematurity, Larissa was entitled to the City of Newton's Early Intervention Program services, so twice a week an occupational therapist came to our house and worked with Larissa on fine-motor tasks and taught Kelly about appropriate games and exercises so she could work with Larissa the rest of the time.

When Larissa was about two, a pediatric orthopedist noticed that Larissa's right leg had some hypertonicity. Lacking the governance of a complete set of neurons from the left side of her brain, the muscles had a tendency to fire unopposed, which led to a generalized tightness. (Corticospinal tract neurons not only stimulate muscle contractions but also regulate reflex motion, and when they are injured the result is spasticity or hypertonicity. This is why children with profound cerebral palsy are contorted by muscles contracting without the brain's regulation.) It was subtle—we had not noticed—but it was clear that the calf muscles of her right leg were pulling her foot down, and the muscles behind her thigh were pulling her leg into a bent position.

"Let's start with an AFO," the orthopedist said, using the acronym for a thin plastic brace that keeps the foot flexed at the ankle. "Depending on how she develops, there are other things we can do."

"Like what?" I asked.

"We've seen a lot of success with Botox," he said, referring to botulinum toxin, the same thing that is used to make wrinkles go away temporarily. In kids with hypertonicity, the Botox is injected into the muscles that are overfiring, causing a partial paralysis that allows the other muscles to compensate. "The downside is that, with Botox, we aren't really training the brain to compensate for this hypertonicity; we're just shutting down the ability of those muscles to contract."

"What are our other options?"

"Well, we can always do a tendon-release surgery." This procedure lengthens the tendon connecting the muscle to the bone, compensating for the uncontrolled contraction of the muscle that has shortened its length and caused the foot to tip downward.

We went down to the brace shop in the basement of Children's

Hospital where Larissa's foot and leg were measured, and when we returned three weeks later, Larissa was fitted with a thin plastic device, the AFO, that hugged the back of her calf, the back of her ankle, and the underside of her foot, with straps coming across the front; it was decorated with pictures of Tweety Bird. Her foot still fit inside her sneaker, and its rigidity prevented her foot from tipping down past the flattened position. This would be the first-line therapy for her hypertonicity.

Over time, Larissa clarified for us the enormously ill-defined prediction of the neurologists during her first week of life. Her major disability would be her right hand. Once a few months of development had refined her volitional movement, it became clear that "Righty," as her hand was known to her occupational therapists and to us, would open and close only slowly, without precision or strength.

Even now I try to imagine what it must feel like to have to think consciously about opening one's right hand. I have wondered if it's like the sensation I get when I sleep in an awkward position and wake up to find my arm has "fallen asleep" due to a temporary loss of blood flow that prevents me from moving my hand normally. Once I asked Larissa what it felt like, and she gave me a quizzical look. "It feels like Righty," she said.

"It will be an assist," her neurologist said. "But I don't think she will get more than that out of her right hand."

I held Larissa's right hand for moral support while the left arm was wrapped first in cotton and then in dampened fiberglass strips that quickly hardened. Given a choice of turquoise, green, or pink, Larissa, now five, had chosen pink, and a layer of stretchy adherent material was wrapped around the cast, making her left arm extremely bright. Larissa's gaze showed both fascination and trepidation.

Then we were off to our furnished apartment in a comfortable but nondescript apartment complex in Birmingham, Alabama, where we were to stay for a month while Larissa went through a special therapy program. We climbed the steps, already beginning to sweat in the heat of the Birmingham summer. Air-conditioning units buzzed around us, working overtime to keep the apartments cool.

Reggi Lutenbacher, who would be Larissa's occupational therapist for the next month, was close behind us. We helped her bring large bins filled with toys up from her minivan: Connect Four, Yahtzee, Trouble, and other games requiring fine-motor manipulative skills, along with a child-size table and a chair. The living room became a fine-motor therapy space.

Each morning at seven, Reggi would knock on the door. Thirty-two weeks pregnant with her first baby, Reggi had charm, enthusiasm, and an unselfconscious way about her. Smiling and ebullient, always upbeat, Reggi immediately got down to business. "Good morning, Larissa. Would you like to get dressed first or have breakfast first?" The choice didn't matter, because both activities would be converted into a fine-motor-skill-building task with real-world utility.

"Remember, Larissa, elbow down by your side. I want to see thumb and pointer holding on to that knob. Now pull, Larissa. Squeeze hard and pull. I'll be your left hand over here on the other knob." She compensated for Larissa's left hand, which was casted and inaccessible.

"Good job, Larissa" was a constant refrain, heard at least a dozen times each hour.

Tasks were discrete—dressing, brushing teeth, playing games—and never lasted more than twenty minutes; the motions were repetitive, designed to improve range of motion, strength, and dexterity of Larissa's right hand.

"Good, Larissa," Reggi cooed. "Now with the next grape, remember, elbow at your side, get the grape between thumb and pointer, and then turn your hand over. I want you to see that dot each time before you put the grape in your mouth." Reggi had painted a dot on Larissa's wrist just up from her pinkie finger, and the dot could only be seen when Larissa completely rotated her forearm, against the muscle tightness caused by her brain injury, so that her palm faced the sky.

Our prior experience with physical and occupational therapy were the sixty-minute sessions twice a week at Children's Hospital. Longer sessions weren't considered—no one really thought a five-year-old could tolerate more than an hour at a time without becoming restless.

"Have you had enough of Connect Four, Larissa?" Reggi asked when Larissa started to fidget and lose focus.

"Yes, Reggi. Can we do something else?"

"Let's do our weight-bearing exercises, Larissa. But first, let's clean up Connect Four." As Larissa and Reggi put the pieces back in the box, Reggi gently cued her. "Remember, Larissa, elbow by your side."

We were in Birmingham for Larissa to participate in the Constraint-Induced Therapy program at the University of Alabama at Birmingham, a seriously intense and uniquely effective therapy program that was the result of research conducted by iconoclastic psychologist Edward Taub.

Taub's original research in macaque monkeys grew out of a surprising observation: If the afferent, or sensory, nerves of a monkey's arms are cut, leaving the monkey without sensation in both arms, the monkey learns—over a period of days—to cope. Movements are somewhat more clumsy than they were prior to the loss of sensation, or deafferentation, but the monkey quickly relearns how to do everything, from swinging on branches to feeding and grooming. However, if the afferent nerves are cut on one side only, the monkey stops using that arm entirely.

Scientists were puzzled. Neuroplasticity was obviously working for these monkeys—they quickly compensated for the loss of sensory stimuli when both arms lost sensation and they had no choice. They used other sources of information—particularly their vision—to help their arms accomplish necessary tasks. So why didn't neuroplasticity help a single deafferented limb regain function?

Taub's theory was that a process of learned nonuse was at work. After one-sided deafferentation, it was easier for the monkey to learn to accomplish tasks with its unscathed arm then it was for it to learn how to use the arm lacking sensation. In essence, the monkey learned not to use the damaged arm.

Working in his laboratory in Silver Spring, Maryland, Taub took an adult monkey who had had a one-sided deafferentation and was getting along nicely with its other hand, entirely ignoring the deafferented limb, and he put the good limb in a cast so the monkey couldn't use it at all.

Nearly immediately, the monkey began to use its deafferented arm, presumably because it had no choice. Neuroplasticity kicked in, overcame the learned nonuse, and within hours, the monkey was accomplishing routine tasks with the deafferented arm. In subsequent experiments, Taub showed that if the cast was removed after a couple of days, the monkey went back to using the normal limb, and the deafferented limb was once again neglected. However, if the cast was left on for five or six days, the monkey regained function of its deafferented limb and continued to use it even after the cast was removed from its normal arm. The deafferented limb never became normal, and was always slightly more clumsy than the unscathed limb, but the regained function was significantly impressive nonetheless. The monkeys' ages didn't seem to matter, although Taub wondered whether younger monkeys would demonstrate even more plasticity than the older ones.

Taub thought he had a treatment that would work with stroke victims, but just as he was about to move his research into humans, politics intervened. Alex Pacheco, a graduate student at nearby George Washington University, volunteered to work in Taub's lab caring for the monkeys. Unbeknownst to Taub, Pacheco had recently helped to found People for the Ethical Treatment of Animals (PETA), a militant animal rights group. Pacheco was a spy.

Pacheco took photos of the animals and secretly invited sympathetic veterinarians to visit in the dead of night. Finally, he called the state police to report cruelty to animals. The police raided the monkey facility, impounded the animals, and arrested Taub, charging him with 119 counts of cruelty to animals. A media storm erupted, the National Institutes of Health canceled Taub's grants, and the scientific community split, with loud advocates of both animal rights and scientific research publishing letters to the editor and making speeches.

At trial, it came out that Pacheco had probably created the unsanitary conditions he reported by not cleaning the cages and not caring for the animals during a two-week period when Taub was on vacation; it was also found that some of the photos he had submitted to the police and subsequently leaked to the media were doctored. Of the original 119 counts, 113 were dismissed, and the other

6 were dismissed on appeal. Eventually, the NIH reinstated Taub's grants, but the damage to the research had been done.

In an interesting coda to the story, the monkeys became the subject of a multi-party custody battle, and most of them ended up being euthanized. At autopsy, pathology studies of the brains demonstrated significant evidence of the neuroplasticity that Taub had anticipated, providing important evidence that plasticity exists in adult mammals.

In 1986, five years after the police raid, Taub was offered a grant and a position at the University of Alabama at Birmingham, and he reopened his research there, this time focusing on adult stroke victims.

The protocol was similar to the monkey experiment. Each stroke victim signed a contract stating he would keep his unaffected arm in a sling for greater than 90 percent of his waking hours. For six hours each day, the stroke victims participated in shaping activities, a special type of therapy designed to increase a limb's function by breaking down the desired activity into its component parts and then building the activity up, one step at a time.

Taub's stroke victims developed connections too, although the anatomy of the connections (was it spinal cord or cortex?) was not entirely understood.

His papers were groundbreaking. Practitioners in the field of neurology had believed that function lost to stroke was gone forever, but here were Taub's research subjects regaining significant function. (Taub compared his subjects with a group of stroke victims who'd received conventional therapy and demonstrated what was well known: traditional therapy did little for these disabled men and women.)

Applying the same principles of constraining the adapted behavior and working on the lost function, Taub demonstrated the effectiveness of his program for stroke victims, victims of traumatic brain injury, patients with multiple sclerosis, and even those with a loss of speech. For these last individuals, he insisted that they not use the mechanisms they had developed to compensate for their inability to speak: signing, motioning, writing notes. Then he developed a set of playing cards with a series of pictures that required increasingly complex speech to articulate and had the research subjects play Go Fish.

In 1995, he published a paper suggesting that his constraint-induced therapy program might be useful for children with cerebral palsy or very early stroke. He wondered whether the significant neuroplasticity present in children might make the program even more successful than it had been in adults.

A doctoral research assistant, Stephanie DeLuca, took on the project, and they set about designing a research protocol. However, the children's hospital at UAB would not permit the children to have their unaffected arms restrained around the clock, and the two sides were at an impasse; the hospital called full-time restraint unethical, and Taub refused to run the program with only part-time restraint for fear that doing so would prevent the children from unlearning their learned nonuse that was a foundational principle of his work.

Eventually, a respected therapist from an outpatient therapy program nearby offered to supervise the constraint, and the program began: four weeks of constraint, with intensive therapy using the shaping methodology for six hours each day and a constant focus on useful tasks.

"Parents will say to us, 'There is no way my child can do six hours of therapy a day,'" said Dr. DeLuca, who now runs the program that Larissa was in. "We don't accept that." What she found was that parents had more trouble watching their children struggle—even if the children ultimately succeeded—than the children had doing it.

The results were relatively profound: compared to children who participated in traditional therapy, children who participated in the CI program showed a greater gain in function that lasted for a longer period of time. Most meaningful to Taub and his associate was that the children in the program and their parents reported that they were able to do more things that improved their quality of life than they had been able to accomplish previously.

Hour after hour, in our temporary Birmingham living room, Larissa and Reggi worked on the same exercises, strengthening muscles, trying again and again to accomplish the pinching, grasping, reaching, and turning that would recruit neurons to the task of expanding the strength, range, and dexterity of her right hand.

We quickly adopted Reggi's phrases: "Whoa, Joe" when Larissa did something particularly impressive, like carrying a cup of water across the room. Or "Now, Larissa, you're all catywhompus" when Larissa took a disorganized approach to opening the refrigerator door instead of doing as Reggi had shown her: lining herself squarely up in front of the door, bringing her elbow to her side, rotating her arm, and grasping the door handle.

By far the most skilled occupational therapist Kelly or I had ever met, Reggi never tired, never lost her patience, and never had a situation she couldn't redirect into an activity that enhanced the function of Larissa's right hand.

Each day, after six hours without a break, during which time Larissa dressed, made breakfast and lunch, played countless games, and did strengthening exercises, Reggi said good-bye until the next morning.

A month later the cast came off, and the effect of neuroplasticity was apparent: Larissa's right hand could rotate further, lift more, and operate with more precision than it had when we'd arrived in the stifling heat of Birmingham.

"The way that neuroplasticity occurs is with repetition," explained Dr. DeLuca. "And through repetition, the activity becomes permanent. That's how the brain works."

Sarah Habib, the New Hampshire preemie who had survived against all odds, also went home, a hundred days after her precipitous birth. In many ways she was like any other newborn. But in her case, she didn't grow. At a year, she was barely ten pounds, and while most children at that age are thinking about walking or trying to, Sarah had difficulty sitting up.

Sounds came too—all babies play with their voices—but most babies' sounds turn into imitations and then coalesce into repetition and words; Sarah's didn't change. Sometimes it seemed like she was trying but just couldn't form the words. Was it a motor problem? Did the muscles that control her voice lack coordination due to her brain injury? Or was it a cortical abnormality in one of the speech centers of the brain? It was so difficult to sort out at such an early age.

By the time Sarah was two, Kim and David were beginning to understand the consequences of her injuries: cerebral palsy affected movement on both sides of her body, her speech, and her ability to eat (at about that time she had a feeding tube surgically placed in her stomach). Sarah's vision was meager, and she had only a couple of words.

Sarah received the same set of services that Larissa did—occupational therapy, physical therapy—and she had innumerable visits with doctors, from orthopedists to neurologists. Kim also had confidence in alternative therapies. While Sarah was still in the NICU, a good friend who was a Reiki master came and performed a Reiki ceremony on Sarah to channel good energy to the tiny child in her incubator.

Once home, Sarah continued to receive alternative therapy. For a while Kim took her regularly to western Massachusetts to visit a practitioner there who performed craniosacral therapy to get energy to flow between the brain and the rest of Sarah's body. Closer to home, she had visits with another practitioner who worked with her fascia and viscera to help "realign" her. "We were just trying to see if there was anything we could do," Kim said.

When I visited recently, Sarah was dropped off by the school bus and came in the door, taking deliberate, stilted steps. Now seven, she had walked without help for just over a year. A smiling and happy child with short brown hair, Sarah had made significant gains.

"Did you have a good day at school?" David asked.

"Yes," Sarah said, drawing out the final *s*.

"Did you do math today?"

"Yes."

"Did you count by ones?"

"Yes."

"What other numbers did you count by?"

"Yes."

Sarah had recently gotten a communication device that spoke when she pushed a button on a console coded with bright images of everyday objects. Kim and David hoped this would help her "speak," but they realized that although the device would help Sarah overcome the difficulty of speech, it wouldn't help with the challenges of language.

Because Sarah's vision was limited and her ability to speak complicated by her cerebral palsy, Kim wasn't entirely sure of the extent of her daughter's intellectual capacity or of her specific limitations. Sarah knew all of her letters and numbers, but Kim admitted, "Cognitively, it's hard to tell where she's at."

In the mainstream school where Sarah went to first grade, she had a full-time one-on-one aide, but sometimes the help Sarah got was baffling. Schoolwork came home printed in a tiny font obviously too small for Sarah to see. Drawings showed evidence of the aide's coloring, not necessarily Sarah's, and Kim heard that Sarah wasn't allowed to move around the classroom or interact with the other children—the school administrators were afraid she would fall and hurt herself.

Relatively insulated from the outside world, Kim saw significant progress on a daily and weekly basis as her daughter gained strength and skills and learned new things. "It's only when you see her with other children her own age that you realize how far behind she is," Kim said.

What's entirely impossible to understand is how Sarah perceives the world around her. "We were on a train at this park up in Vermont," recounted David. "Sarah likes trains a lot, and she started making this sound that she often makes, 'Aah-aah,' while rocking her head back and forth. It's just something that she does." A father sitting nearby with his kids picked his kids up and moved them to the other side of the train, as far from Sarah as he could get.

"Sarah doesn't notice," David said. "At least I don't think she does."

David worries about what will happen when Sarah is older and notices that people move away from her or don't want to play with her. "Right now, at her developmental stage, it's not a big deal. But as she gets older, the impact on her self-esteem worries me," he said. Of course one of the problems is that they don't know exactly where Sarah is developmentally.

"We haven't had her tested," Kim said. "She knows her address, and she can count to thirty."

"She can count by tens to a hundred," David offered.

"I'd say she's probably about four or five," Kim guessed. "But she doesn't grasp cause and effect," she added. "If she wants to pick something up, she often doesn't realize she has to put what's in her hand down to pick up the next thing. In some areas she's very advanced, but in other areas she's very far behind."

Sarah came into the kitchen. "Du," she said, her way of saying that the television show was over. "Ju?" she asked. "Ju?" And Kim made her a sippy cup of juice.

"You did a good job," she told Sarah.

CHAPTER 11

The Plasticity Treadmill

Jason Carmel took over a lab bench in Jack Martin's lab, a warren of small rooms packed with equipment on the ninth floor of an aging City College building on the West Side of New York City. Looking through dirty panes, Carmel could see across Manhattan to the East River and Queens. He was working on the building blocks of his theory: well-defined animal experiments that would prove—or refute—his overall hypothesis that same-side neurons could develop into a viable system of control for parts of the body abandoned by injured neurons. What if, Carmel thought, he could reverse the process that led to opposite-side control and recruit the frail same-side neurons to action? Could the right brain be convinced to control both the right and left sides of the body after an injury to the left brain? In Larissa's case, could her right brain be made to control her right side after her left-brain injury at birth had abandoned the right side of her body?

The theory was radical because it contradicted conventional wisdom about treating stroke and cerebral palsy, which focused on rehabilitating injured brain, not recruiting healthy brain to take on a new and unexpected task.

Although he didn't know Larissa, Carmel's idea would translate to her situation like this: because of the injury to the left side of

Larissa's brain, the dominant corticospinal tract fibers that typically control the right side of the body are underdeveloped, and this may have allowed greater-than-normal development of those few corticospinal fibers that ran from the right side of Larissa's brain to the right side of her body. Maybe, thought Carmel, he could harness the innate plasticity of these same-side neurons and coax extraordinary function out of their narrow axons.

Ignoring decades of research that held that the best way to help a patient regain function after a brain injury was to focus on rehabilitation of the injured brain cells, Carmel decided to see if a bold approach leveraging an arsenal of therapies could teach the uninjured side of the brain to control both sides of the body. "Our goal is to drive bilateral control from one side of the cortex," Carmel explained, "by activating the long-dormant ipsilateral [same-side] fibers."

The potential side effects of this approach were not trivial. Patients with spinal cord injuries worked hard to strengthen and regenerate synapses only to develop excruciating pain syndromes linked to the new connections they had formed. A few researchers even thought that after a stroke, developing the same-side neurons actually impeded the rehabilitation of the injured brain cells and the relearning of the cells with connections that crossed over.

The biggest uncertainty, from the perspective of Carmel and his new research mentor Jack Martin, was the extraordinary complexity of the brain itself. It's one thing to stimulate the brain of a rat in a particularly well-known region of the motor cortex and observe the expected twitching of the paw, but that doesn't mean that there aren't other ways for a signal to go from the motor cortex to the paw, or that a host of other circuits don't modify and influence the action between the initiating cell in the motor cortex and the muscle fiber. With his years of experience methodically piecing together the anatomy and function of the corticospinal tract, Martin cautioned Carmel that circuits they hadn't yet considered could torpedo the ambitious plan.

While building the foundation of a research program in Jack Martin's lab, Jason and Amanda Carmel had bought a condominium in a newly renovated building in South Harlem, close to work and to Central Park, and they'd started a family. Like any dual-career couple, they struggled to balance their jobs with their family's needs, and except for their own sleep requirements, they managed to meet the needs of their family and their jobs.

One hundred blocks to the south, David had returned from the West Coast and had also married and had children. The Carmel brothers got together regularly, now with their small children in tow.

On a recent winter night, Jason Carmel finished up an experiment, took the rats back to the animal facility, bundled up against the cold, and made his way out into the New York twilight. It was a fifteen-minute walk to his condo, dodging road construction and passing low-rise housing developments built in the 1950s.

His apartment building appealed to young families, and evidence of children was everywhere; strollers and trikes competed for space. He went up an elevator and turned his key in the lock, and the warmth of home spilled into the hallway. Two children came running, a third crawling, and their nanny smiled in the background. In the suburbs there would be space for a playroom, but in the city, it was the living room that had a miniature play structure and the remnants of a pillow-and-blanket fort. Amanda's clinic ran late that evening, so Carmel was on his own for a while. Sometimes he marveled at the innate ability of his children, like most children, to learn complex tasks that combine curiosity with intricate movements. Most of the time, he left work behind and just enjoyed the simple, satisfying time with his kids.

In the lab, the tools available to Carmel were somewhat crude: electrical stimulation; systems of restraint, like the cast Larissa wore for a month in Birmingham; and planned activity for the impaired limb. Success would depend on two things: the validity of Carmel's hypothesis and his creativity with the tools available to him.

With some funding from Martin, and later a small grant from the foundation set up by actor Christopher Reeve after he suffered a catastrophic spinal cord injury while horseback riding, Carmel set up his experiments and got to work.

Because in science nothing can be taken for granted, Carmel's initial experiment had three objectives. First, confirm what everyone knows is true—that after an injury to the corticospinal tract, a rat can be taught to regain much of the motor control it lost through injury. Second, explain the circuits of neurons that allow for recovery of function—in essence, understand the anatomy of plasticity. And third, use activity to strengthen those circuits.

Carmel's rats were white and six inches long, and they had beady red eyes. It turned out that their eyesight was very poor, so they depended on the sensory input from their whiskers to navigate the world.

For this experiment, the rats spent eight weeks learning to walk across a ladder. Carmel fashioned the ladder with high walls made of clear Plexiglas (so Carmel could watch the rat on the ladder, but the rat couldn't jump off the side) and rungs that could be moved or taken out so that they were unpredictable—the rat had to use its eyesight and whiskers to sense where it was and to put its paws down in the correct place, not just memorize the distance between regularly spaced rungs.

For the rat, training involved being picked up and put on a short chute that led to the ladder and then running across to the other side, where a Q-tip dipped in sugar water awaited as a treat. Next to the ladder, a video camera recorded it all.

Carmel's assistant, Lauren, who had had a career as a personal injury lawyer before taking a job in the lab so she could work her way through her pre-med courses, had become something of a professional rat trainer. A rat ran ten times one way across the ladder, and then ten times in the opposite direction. Then Lauren changed the rung pattern and started over.

Rewarded after each trip across the ladder, the rat was cooperative and fast, its feet scurrying across the twenty rungs with seemingly perfect accuracy.

Halfway through one set of exercises she called to another research assistant in the next room. "Can you get me another Q-tip? The rat ate this one."

When the training exercise was over, she stopped the video and downloaded the data into the iMovie software on her Macintosh computer. In real time, the rat moved much too quickly for Lauren to assess the accuracy of its paws on the rungs of the ladder, which was the purpose of this exercise. But looking frame by frame on the computer, Lauren could assess each step and determine whether the rat put its paw down directly on the rung or overstepped slightly so the rung hit the wrist. Understeps were easy to detect, as the paw missed the rung entirely and the rat had to catch itself so it didn't stumble. By the eighth week, the rats ran across the ladder flawlessly.

Then it was time for surgery. Anesthesia was injected, and the skin at the front of the neck incised. Looking through a microscope and using a fine set of scissors, Carmel dissected the neck, gently pushing aside the trachea and passing between the carotid arteries until he found the base of the skull. He drilled a tiny hole in the skull and then used the scissors to slice through the dura mater that formed the protective layer around the brain and spinal cord; he identified the dual pyramids of the corticospinal tract just below the area where they crossed and cut through the left pyramid, severing connections between the right motor cortex and the left side of the body. Then he closed the skin with tiny surgical staples and waited for the rat to wake up.

A day or two later the testing began. The rat was run across the ladder just as before—ten times in one direction; ten times in the reverse direction. Lauren could see the impairment just by watching the rat, but when she sat down at her Mac and slowed the video, it was really apparent: the left foot kept missing the rungs—overstepping and landing on the wrist, or understepping and stumbling.

But over several days, the left paw got markedly better. The rat was never as accurate with its left foot as it was with its unimpaired right, but the left paw regained significant accuracy.

The challenge for Carmel was understanding the mechanism of learning that allowed the rat to improve its ladder run after the injury to its corticospinal tract. Prior research had demonstrated that once severed, the corticospinal tract did not repair itself. So did the same-side tract rise to the occasion, as Carmel had hoped, or was it some other mechanism at work? It was time to find out.

Back under the dissecting microscope, Carmel consulted an atlas of rat-brain anatomy to make sure he was in precisely the right spot. He identified the motor cortex on the rat's right side—the side that had been disconnected from the left side of the body when the corticospinal tract was severed weeks earlier.

Now, he injected a tiny amount of a special tracer into the cortex. He finished the surgery: cement went on in place of the skull and skin, and the rat was again awakened and allowed to recover before facing the ladder in the days that followed.

Unbeknownst to the rat, the injected dye—which traveled from cell body to axon only and never in the reverse direction—was making its way down the intact corticospinal neurons. Over a period of several days, the dye came to occupy the nerve axons of intact corticospinal tract neurons.

After the tracer dye had been given enough time to fully color the intact neurons (the dye could not cross the site of injury, so injured neurons would not absorb the dye), training ended, and it was time to see if Carmel's hypothesis was right.

After a purposeful overdose of anesthesia, Carmel made cross-sections of spinal cord a few microns thick and affixed the slices to slides that he could examine under the microscope.

Using a chemical that turned the axons black, Carmel looked for the tracer that marked the axons of corticospinal tract neurons. Bright black dots for axons passing perpendicularly through the slice on the slide; short lines for axons traveling parallel along the plane of the slice.

He compared the results to a set of spinal cord slices made from control rats, those who hadn't participated in the training regimen.

In the butterfly-shaped cross-section of the spinal cord, Carmel saw a collection of black on the impaired side in the middle—much more dense than what was seen on the brain slices of the control animals. These were the same-side corticospinal tract connections. And on the unimpaired side, he saw nothing. His hypothesis was correct: by the process of plasticity, these same-side neurons had come to the rescue of the obliterated neurons.

This was good news, but hardly a breakthrough. Anyone with a brain injury will say that although some skills are regained over time, physical and occupational therapy alone—the human equivalent of ladder running—does not restore normal function. Given what Carmel knew about the significant improvement in his rats' function but the only minimal improvement in the function of humans who had the equivalent injury, there was more research to be done.

He embarked on a series of experiments to see whether he could enhance the process of plasticity. By infusing a temporary paralytic into the area over the motor cortex, he learned about the long-term effects of brain injury on development—those rats who were temporarily paralyzed early in life never caught up, even after the paralytic infusion ended.

In another set of experiments, he implanted a tiny electrode directly into the rat's left motor cortex, the side of the brain that remained connected to the right side of the body through the neurons that crossed over. Peering through the dissecting microscope and watching for the front paw to respond when a tiny stimulation was given through the electrode, Carmel pinpointed the exact location where the electrode should sit. This experiment was somewhat counterintuitive, given the years of research that focused on using electrical stimulation to the injured side (in this case, the rat's right motor cortex that had been disconnected from the rat's left side).

After the electrode was implanted, Carmel secured it with the cement, and over a period of weeks, he provided a low-level stimulation while the training with the ladder was ongoing.

Finally, Carmel was getting impressive results: the density of the same-side neurons labeled with the tracer dye increased with each new tactic. Other researchers had shown that exercise was better than no exercise. Carmel now demonstrated that electrical stimulation was better than no stimulation. And electrical stimulation in addition to exercise was best of all. And of critical significance to Carmel and Martin, who had spent so much time understanding the circuitry, when they looked at slices of spinal cord under the microscope, the stimulation created connections that explained the improved function they saw on the ladder.

Martin and Carmel were proving their hypothesis, at least in rats, experiment by experiment. Injury reignited competition between the injured side of the brain and the intact side of the brain for control of the abandoned side of the body, and the two neuroscientists had proven that exercise and electrical stimulation could help the same-side fibers outcompete for control. They had ladder-running data to prove it, and they had fluorescent Spaceballs data from the slices of spinal cord.[1]

Carmel and Martin often wondered if their data was too good to be true. If they were on to something of significant clinical importance, where was the recognition? Where were the competitors? "Why don't we have a horde of neurologists knocking on our door every day saying, 'We should try this in kids'?" Martin asked.

But they didn't; the clinical community largely ignored their work, and Carmel and Martin were left to envisage the clinical trials they would do in the future. They vowed to do them, even if they had to do them alone.

There were bad days. Rats are great for research—they grow to maturity quickly, have most of the same organ systems as humans, and are cheap—but the translation only goes so far, and that recognition makes animal research a lonely endeavor. Although he refused to see it as a bad omen, as his research with rats intensified Carmel developed a brutal allergy to the rats; his arms broke out in angry hives and his nose began to run every time he handled the creatures. (*At least I'm not allergic to children,* Carmel thought morosely.) Most days, however, were exciting. The research was fascinating, and so far they had found none of the hypothesis-killing discoveries Jack Martin had worried about.

CHAPTER 12

Living the Dream

Predictably, when Larissa got her left hand back after her cast was taken off, she went back to using it, and her impaired right hand went from starring in the Birmingham hand camp to playing a decidedly supporting role. It wasn't that her right hand couldn't reach up and open the refrigerator door; it was just so much easier for her left hand. Testing at the end of the monthlong session showed she had made enormous gains, but we were warned that unless we kept up with her home program—a set of activities designed to reinforce and strengthen the progress she had made—those gains would be lost, at least in part.

Kelly and I made a modest effort to keep up with the home program, but within a month, it was forgotten amid the demands of getting the kids to school and then bathed and to bed in the evening. It is unrealistic to expect a five-year-old to remember to use her right hand for certain activities when it is simpler to use her left, and so Righty didn't receive the attention it deserved.

Three years later, in the summer of 2010, we returned to the heat of Birmingham to do the program again. Our furnished apartment, in a completely different complex, looked pretty much the same as the last one. We were lucky enough to get Reggi back. She was even pregnant again.

"Your right hand has really improved," the therapist who did Larissa's intake evaluation told her.

"We didn't do a great job of keeping up with the home program," I volunteered.

"One way or another, her right hand has gained a lot of ground," the therapist said.

The other difference was that Larissa was now eight and was able to articulate her own goals for the program. She wanted to learn to tie her shoes because she found it embarrassing to be the only one in her class who had to ask for help, and she wanted to learn to zipper her coat—again, this skill would grant her another degree of independence. She also wanted to learn to ride a bike.

A therapist near our home in Newton had started teaching Larissa to type; the question was whether she should be taught to use two or three fingers of her right hand or write it off entirely and be taught one-handed typing. Reggi disagreed with both suggestions.

"I think she can use all of her fingers," Reggi said, and over the four weeks, she showed Larissa how those fingers of her right hand could learn to independently strike the keys of both a computer and a pink Barbie piano.

She concentrated intensely and learned during the course of two weeks to pick up a Yahtzee die between her thumb and index finger, turn her hand over, and drop the cube into the palm of her hand. I told her how impressed I was.

"Great job," I said as I watched Larissa struggle (successfully) to pedal her bike up a shallow hill in the parking lot outside our apartment. "Super work."

A year earlier Kelly had gone back to work, so she and I split the month—I spent the first two weeks in Birmingham with Larissa; Kelly was there for weeks three and four.

"I never hear her complain. Never," Kelly told me on the phone from Birmingham one night after Larissa had gone to bed. "And she tries so hard to do whatever Reggi tells her. But I have to tell you that it is very painful to watch her work so hard to do with her right hand what most children take for granted."

Kelly thought back over the years—over the hundreds of doctors visits and the thousands of exercises with her physical therapists that were invariably scheduled for after school, when most kids were playing with friends or running around in the backyard.

"She inspires me," Kelly said. "She really does, because she's always so positive and never feels sorry for herself. But it makes me feel bad for her."

While Larissa was with Reggi, and after Larissa went to bed at night, Kelly worked on her dissertation—now entering its ninth year in process. Kelly was finally confident she would finish it.

"It's not that I regret one minute of it," she had said several times over the years. "I'd make the same choice again."

Measured by any metric, Kelly has succeeded. Larissa is a bright child, verbal and funny, and she's a hardworking student who attends the same mainstream school as her sister Hannah. She read early, spoke early, and, like many of her classmates, has some difficulty with her handwriting.

To be fair, she seems to have some problems coordinating her perception of objects with her ability to draw, and her paintings and drawings resemble what you might expect from a child two or three years younger. But artistry aside, she stands out for her humor and her personality, not for her right hand. In the indignities of her disability—shoe-tying, basketball shooting, monkey-bar scooting, painting—Larissa is reminded endlessly about what she can't do. But in the context of everything she can do, these issues fade into the background.

Science is incremental, and from the perspective of a career, success can be measured one paper at a time. Jason Carmel was publishing, and after finishing his residency he found himself a full-time research position—another sign of career success.

It seemed appropriate that Carmel's research had outgrown his brother's injury—the most obvious disease target was now cerebral palsy, although the approach of using stimulation to strengthen brain-spinal connections might have a role in the treatment of spinal cord injury. Carmel had come to terms with the fact that his research would not cure David's inability to move or feel below his chest, and during the years since his injury, David had come to terms with this too.

But at times, when the welts rose and the rats misbehaved, Carmel was sobered by the recognition that he was modeling cerebral palsy and spinal cord injury in a six-inch rodent with a wire stuck into its brain through some modeling cement. So many questions remained unanswered, beginning with whether the rat was even an appropriate animal model for the human diseases Carmel really wanted to treat. Humans rely on their corticospinal tracts much more than rats do, and it would take studies in humans, perhaps years in the future, to determine exactly how well these exciting animal-model results translate into children like Larissa.

It was probably impractical to implant an electrode in a child's brain, so Carmel focused on finding a safe and noninvasive way to stimulate neurons. The best option at that point seemed to be transcranial magnetic stimulation (TMS), a cousin-technology to magnetic resonance imaging (MRI) that uses mild magnetic fields to create gentle electrical currents that can stimulate the brain. Best of all, TMS had an established track record in pediatrics.

"I don't think we have any illusions that we're going to take a kid with severe hemiparesis or quadriparesis and get them playing on the high school basketball team," Carmel said. "But we could make some important contributions to the quality of these kids' lives."

Science, increasingly, was a team sport, and Carmel had joined colleagues who shared his vision of therapy that combined stimulation of the brain with training of the injured limb. Carmel focused on direct stimulation of the brain, but Martin had shown that exercise of the disconnected limb was critical, so Carmel teamed up Andy Gordon, a scientist who had a training program for children with cerebral palsy that combined aspects of the Birmingham system—he used a sling instead of a cast to immobilize the unaffected arm—with two-handed activities.

Together, Carmel, Gordon, and Kathleen Friel, another colleague who was pioneering research using MRI to pinpoint the exact extent of brain injury in children with cerebral palsy, were bullish about their prospects of making the first significant advance in therapy for cerebral palsy in a generation. "We have the tools, and we understand the circuitry," said Carmel. "And because the therapies—TMS and constraint therapy—are known to be safe, what's really exciting is that we can get going now."

Carmel planned a clinic focused on cerebral palsy and other neurologic disorders unlike anything that exists today. A kid like Larissa might be evaluated soon after going home from the hospital using Friel's advanced MRI sequences to pinpoint the location and extent of her injury. Then she might come in each week, strap into a TMS helmet, and spend a couple of hours training to do specific tasks while targeted magnetic waves create precise electrical currents that stimulate neurons on the right side of her brain, making the learning process easier, faster, and more effective. Because TMS may work even without exercise, it might be possible to start TMS much earlier than exercise therapy—which requires the cooperation of the child—can begin.

"Given the cat studies and what we know about the human corticospinal system as a bilateral system early on that starts to prune itself back to a single-sided system, our ability to use electrical stimulation during that early time when kids cannot participate in physical therapy could be very, very important," Carmel said enthusiastically.

However, will noninvasive stimulation be as effective as the implanted wire that will never be tried in humans? Will the window of plasticity in children stay open long enough that kids will benefit from these treatments? "I'm never going to be able to treat a newborn baby with an intraventricular hemorrhage, or even an adult in the first week after they injure their spine," Carmel acknowledged. "Will this work when children are two, or twelve? As a neurologist, these are incredibly important questions."

Are there side effects that won't be discovered until human studies begin? Research is rife with great ideas that had unanticipated consequences once studies moved from rodents to humans. But it's hard not to get really excited.

"When we first do TMS, we are expecting modest effects," Carmel said, downplaying his hopes. "But as we better understand the circuitry, and understand which kids respond, we expect the efficacy will go up, and any chance of side effects will go down. Even early on we expect to make meaningful contributions to people's recovery."

Sarah Habib isn't much of a movie fan, but she loves a musical with a melodic score. When a movie becomes overstimulating or her surroundings create too much visual input, Sarah simply takes off her glasses. It's a way to keep the world out, a strikingly effective way of reducing stimulation.

It eludes her parents just how far these coping mechanisms go. Sarah appears to be unencumbered with disappointment that she can't climb play structures like her contemporaries and doesn't have playdates. Is she aware of how different she is from the other first-graders? Are her cognitive limitations such that she doesn't even notice? Unlike Larissa, who discusses right-sided disability and also offers myriad clues to her awareness of her difference, Sarah's thoughts are obscure to her parents and to the other children and adults who try to interact with her.

"How do you know?" Kim asks. "If Sarah hears me say she can't do something, then of course she's not going to be able to do that."

"For me, it was about three years before I really understood that, okay, there are some things she is never going to be able to do," acknowledges David.

He would never admit to being disappointed—it would sound like a betrayal of his daughter—but David is matter-of-fact. "I've never been a parent in any other way," he says. "I adapt to what she needs."

Kim finishes his thought. "We just say, 'How can the situation be the best that it can be?' "

Down on the floor of the living room, empty of furniture that could get in the way of Sarah's play or practice walking, David sits, legs spread in front of him, across from Sarah, who has gotten him to play her favorite game: ball. The blue rubber ball flies back and forth. David rolls it across the floor to Sarah, who stops it, picks it up, and heaves it toward David, sometimes with a bounce, in his general direction. It's a game she could play forever.

"Ball," Sarah says.

"Ball," David agrees.

Kim sees incremental progress. "Every week she does something different, I'm telling you. It's a new word, or better drawing.

When people haven't seen her in a while, they're like, 'Wow, she does that now!' And guess what, now she's potty trained."

Kim is always looking for signs of hope, like the advertisement on the subway wall.

There was the time the family and Kim's parents went to an antique store on Sarah's birthday. Kim's mother wanted to buy Sarah a gift, and Kim was all kinds of skeptical about the likelihood of finding an appropriate present for a little girl in an antique store. "My mother was showing her all of these old music boxes, and there was one with a unicorn, a teddy bear, and a star on it." Kim's mother wound it up, opened the top, and played the tune for Sarah. "Sarah signed *I love the song* to my mother," Kim said. "I wasn't paying attention, but my mother brings the music box over to where David and I are standing and is insistent that she's found a gift for Sarah, and goes to purchase it. We get the music box home and Sarah pulls it out and opens the top and it plays 'To dream the impossible dream,' from *Man of La Mancha*.

"You see: Sarah's living the impossible dream."

Without a lot of enthusiasm, Steve Ringer was trying to find a word to describe the current state of newborn medicine, long years after the heady days when his colleagues discovered surfactant. "Elegant," he said. "I think we are trying to make care more elegant."

Hospitals were pouring enormous resources into checklists and other low-tech systems to reduce errors and preventable complications, and Ringer's NICU was in the thick of it, making sure that every time a central vein catheter was placed in one of the babies, a regimented set of guidelines, procedures, and sterilization techniques were carefully followed. The meetings that were needed to get everyone to contribute to the protocol were mind-numbing. Then there were four layers of hospital administration to go through to get approval for the new system. Finally, a skeptical group of residents, nurses, and fellows had to be won over, trained, and policed.

But Ringer had to acknowledge that it worked. The problem of catheter-line infections pretty much disappeared. No one knew exactly how, because, of course, everyone had always used a sterile technique. It was a mystery, but a pleasant one, even if Ringer couldn't take credit for this obvious success. This was an example of elegant medicine—more standardized, evidence-based, methodical, boring.

The paperwork demands of the job increased too as the hospital asked for more documentation, and the health department and other supervisory agencies wanted to track diseases, infections, and outcomes with greater granularity. This part was frustrating.

After the big discoveries were behind them and newborn medicine doctors turned to the critical task of fine-tuning their field, they realized that it was no longer adequate to publish papers reporting the experience of one hospital. Surfactant was so meaningful that it took only a few dozen babies to show that the stuff worked. But it is a truism of statistics that when a therapy's effect is expected to be relatively small, a very large number of participants need to be enrolled in the study. So to prove that, for example, 40 percent oxygen supplementation is better than 50 percent oxygen supplementation, a very large number of babies are needed.

Enter the Vermont Oxford Network, founded in 1988 with the specific purpose of improving the quality and safety of care in the NICU. The idea was simple, brilliant, and oh-so-tedious: All using the same precise criteria, NICUs would collect hundreds of pieces of information on how they cared for their tiny patients and record the outcomes—from infections to intraventricular hemorrhage, from retinopathy of prematurity to death—and send it in to the central office in Burlington. Expert statisticians culled through the data, merged it together, and then sent each participating NICU a confidential report that allowed the unit to compare itself with other units around the country and, eventually, around the world. By 2010, more than eight hundred units participated in the process.

Over time, this cumbersome bureaucratic system transformed care in newborn medicine. It became common practice for units with higher-than-normal rates of respiratory complications to seek advice from units that had the lowest rates of complications. Experts from centers with the best results were invited to visit lower-performing centers, and best practices emerged, were published in scientific journals, and then trickled down across the country and around the world.

It didn't always work. Ringer participated in one program run by the Vermont Oxford Network that identified a package of best practices used by the NICUs with the lowest rates of chronic lung disease and took it to NICUs that had the highest rates of this common complication. When the results were evaluated years later, the group determined that even though the lagging NICUs had succeeded in implementing many of the recommended best practices, this had had no impact on chronic lung disease itself. "Something immeasurable was going on in those units," Ringer said. "And we were as yet too stupid to understand exactly what it was."

But the adaptation had occurred. Eight hundred NICUs had banded together, had realized they needed to share data and learn from one another and that only through cooperation would elegance come to the constant methodical improvement of care for their tiny patients.

Within the walls of Brigham and Women's, Ringer was also working on ways to make the care more elegant. It is well known that breast milk is better for babies than formula. There are all sorts of benefits, from fewer infections to fewer allergies, and it has even been shown—although some experts dispute this fact—that breast-fed babies are smarter.

Breast milk is even more important for premature newborns. Breast-fed premature babies get fewer infections and are less likely to have cardiovascular complications. But the most significant advantage is that breast-fed babies are much less likely to get necrotizing enterocolitis—a devastating complication of prematurity in which the intestines don't receive sufficient oxygen and die. Necrotizing enterocolitis, known as NEC, can sometimes be treated by simply giving the intestines a rest, but it often requires surgery to remove the dead piece of intestine and reconnect the remaining gut. For reasons that aren't entirely understood, breast milk prevents NEC.

For years, the Brigham and Women's NICU, like most NICUs, has had a system to feed babies their mothers' milk. Women were encouraged to pump breast milk at home and freeze it, and the NICU then defrosted the carefully labeled units of milk and dripped them into the babies' stomachs or put it in bottles when the babies were big enough to suck and swallow.

But what to do for babies who didn't, for whatever reason, have access to a steady supply of their mothers' breast milk?

Wet nurses—lactating women hired to provide breast milk to other babies—have an important role in history going back millennia. But their popularity declined markedly in the early twentieth century with the recognition that some diseases could be transmitted via breast milk and with the development of reasonably safe infant formulas. Human milk banks—along the lines of blood banks—developed in Europe and the United States in the early 1900s to provide for babies, particularly premature babies, when their mothers' milk was unavailable. With the advent of infant formula, milk banks moved toward the fringe of child-rearing, as medical professionals advocated formula over breast milk. The AIDS crisis in the 1980s, and the discovery that the virus could be transmitted in breast milk, caused most milk banks to close.

But as evidence mounted that breast milk was beneficial, milk banks learned from their colleagues in the blood-bank industry and developed highly accurate systems to detect disease in their banked milk and to pasteurize their product.

When one mother in the Brigham and Women's NICU asked her nurse if she could arrange for her baby to receive donor breast milk, the NICU administrators put their heads together quickly and then said yes.

Lactation consultant Tina Steele thought the Brigham and Women's could do even better. She pulled Ringer aside and suggested the NICU provide banked breast milk to all the unit's babies whose mothers wanted it.

"What?" Ringer said. "How would that work? And who would pay for it?"

"I'm not sure," Steele responded. "But I know it would be good for the babies."

"Why don't you put together a proposal," Ringer said. "And make sure it's evidence-based."

A couple of months later, Steele was back with a proposal for how to use donor breast milk in the NICU, a review of the literature, and information on the cost involved.

She presented her ideas to the leadership team Ringer had assembled. No longer was this a group made up entirely of doctors; the new mantra was collaboration, and that meant that leadership involved all of the clinical disciplines, Ringer's team of physicians as well as nurses, respiratory therapists, and the other clinicians that cared for the NICU babies. The group was impressed—the strength of the medical evidence was overwhelming.

They had to find funding—banked milk costs nearly five dollars per ounce—but they identified some philanthropic support and hoped that their study would demonstrate a cost savings elsewhere in the babies' care.

The day before the program was set to begin, a woman with HIV infection delivered a twenty-five-week newborn. The baby was the perfect candidate for the program since she couldn't safely receive breast milk from her mom, so Steele opened the program early.

It was far from a wonder drug like surfactant, which changed newborn medicine overnight, but it was backed by medical evidence and had broad support in his unit, so Ringer thought it might fit into his concept of elegant care.

In 2006, after eighteen years alone at the helm, Ringer hired a younger physician to run the NICU on a day-to-day basis—younger, but not just out of fellowship. This physician spoke about developing cohesion and teamwork. Ringer listened to this and heard his own voice from years earlier. He was stepping back to work on the big picture, although it was often unclear just what that big picture looked like.

Of late, Ringer had seemed distracted and short-tempered. For a guy who was hired in part for his unflappability and his ability to instill calm, it was disturbing to find himself snapping at nurses and even losing his cool and yelling at colleagues. Around him, he heard that people were wondering what he would do next.

During a recent summer, Ringer and his longtime mentor Bill Speck had gone fishing again off the coast of Cape Cod. Speck had retired after a long career as a successful hospital administrator that culminated in a stint as chief of the massive Columbia Presbyterian Medical Center in New York. Ringer, who had once envisioned himself rising to become a hospital president, had an opportunity to reflect on his own career's trajectory and his recognition that, like newborn medicine, his professional passion had evolved and was satisfied by solving discrete problems in the NICU; he didn't need to tackle the broad administrative challenges facing a high-level hospital administrator. Floating with Speck on the Vineyard Sound, nearly thirty years after they first fished those waters together, Ringer looked back over his career with a degree of satisfaction.

The Brigham and Women's NICU was now a place where babies uniformly received the best medicine had to offer. A large staff from different clinical disciplines worked together to constantly improve the care they provided to the babies, meeting regularly to evaluate their work, objectively and critically, but with a newfound level of professionalism. Vermont Oxford provided them with a context, and the esprit de corps that Ringer had built year by year drove the subtle advances that were truly elegant. Ringer had ushered in the use of surfactant and a host of big discoveries over his years in charge, as well as the administrative changes that were often more difficult. Recently there'd been talk about a whole new unit that would be larger and provide more space for families to kangaroo and engage in the care of their babies, and Ringer was looking forward to watching the next generation of advances.

When Larissa's older sister Grace was five, she wanted to learn to ride a bike. We went to a field near our house, I held the bike while she positioned her feet on the pedals, and then I gave her a push.

"Pedal," I yelled. And she did, eventually hitting a bump and falling over into the soft grass. We did the same thing again and again until she no longer fell.

A couple of hours later, with one of Grace's knees bloodied and a huge smile on her face, we went home and she proudly told her mom she knew how to ride a bike.

Back from Birmingham, Larissa and I went to a stretch of gently sloping tarmac behind a local elementary school to complete the bike-riding lessons she had begun with Reggi.

First, I removed the pedals, and she practiced coasting across the pavement. I ran alongside to catch her when she lost her balance.

It was slow going, but she gained confidence in her balance, learning to adjust for the inherent asymmetry in the function of her right and left sides. "Aim for the fence," I said, and at first we made it only a few dozen feet. Frustrated, we took a break, sitting in the air-conditioned car to recover from the sweltering heat reflected off the parking lot.

I'd been inspired by Kelly's gentle perseverance and the methodical way she had taught Larissa to gain developmental skills, and we kept coasting along the pavement until Larissa could, in fact, glide the length of the parking lot and I had to stop her from running into the fence.

The next day we tried something else: this time the pedals stayed on, and the bike was balanced on a special bicycle treadmill so Larissa could learn to keep her feet on the pedals as she powered the wheel around and around. We counted fifty revolutions, then took a break, and then did it again, and again.

On the third day, we moved back to the field and once more picked an area that was gently sloping so we could recruit gravity to our side. Like Grace years earlier, Larissa readied herself atop the bike, I gave her a push, and she managed to pedal a few feet before weaving and falling into my arms. We did this at least twenty times, never making it more than a handful of bike lengths before her right foot came off the pedal or she hit a bump in the grass and tipped over. Frustrated and sweating, we cooled down in the car.

Once more, we set up at the end of the field, and Larissa got ready to pedal. I gave her a push, yelling, "Pedal, pedal," and she pedaled as hard as she could. This time her feet stayed on and she reached a critical speed; running to keep up, I watched her pedal across the field.

We went back to the house, sweating and dusty, and told Kelly about Larissa's cycling achievement, but Larissa's enthusiasm about the accomplishment was tempered. She was proud of herself, but she seemed to recognize how hard it had been to gain this skill that had come so easily to her sisters.

Nine years after that terrifying morning when Larissa was born, Kelly and I share a glance, as we sometimes do when we watch Larissa accomplish one or another improbable task. We each have a slightly different expression, but it is reserved for times like these.

First, there is gratitude that our child is riding a bike; at one point, neither of us knew if she'd ever even walk. Next, there is sadness that so many times every day Larissa is reminded of the activities—trivial and significant—that are harder for her than they are for her friends. Last, we share awe at the spirit and determination that have helped Larissa more than all the miracles of therapy and neuroplasticity combined.

ACKNOWLEDGMENTS

In addition to Kelly, to whom this book is dedicated, I am perpetually inspired by Larissa for her tenacity, strength, and humor, and by her two remarkable sisters, Hannah and Grace.

I am grateful for the sustaining support of family members and friends, including Susan Lynch, who served as Kelly's right-hand mom during Larissa's early years; Kelly's sister, Karen; my parents, Steve and Judy; and my brother, Elias.

I am grateful to Steve Ringer and Jason Carmel for sharing with me their science, their practice of medicine, and their vision for improved pediatric care.

Many friends, colleagues, mentors, and even caregivers of Larissa gave generously of their time and advice, and some reviewed sections of the book. They include Elisa Abdulhayoglu; Craig Conway; Sabrina Craigo; Jon Davis; Stephanie DeLuca; Adre du Plessis; Michael Epstein; Lorraine Figelsky; Sarah, Kim, and David Habib; Gabi Harrison; Joe Kaempf; Reggi Lutenbacher; Jack Martin; Amanda Rodriguez; Saroj Saigal; Jane Stewart; Ed Taub; Linda Van Marter; and Joe Volpe.

Last, I am tremendously appreciative of Helene Atwan, my editor at Beacon Press; Julie Silver at Harvard Health Publications; my literary agent, Linda Konner; and my longtime friend and mentor Evan Thomas, who suggested this book was possible.

NOTES

Chapter 3: Gifted Hands

1. T. Raju, "From Infant Hatcheries to Intensive Care: Some Highlights of the Century of Neonatal Medicine," in *Fanaroff and Martin's Neonatal-Perinatal Medicine,* 8th ed., eds. R. Martin and A. Fanaroff (Philadelphia: Mosby Elsevier, 2006), 3–18.

2. Ibid.

3. Tetsuro Fujiwara et al., "Artificial Surfactant Therapy in Hyaline-Membrane Disease," *Lancet* 315 (1980): 55–59.

4. L. O. Lubchenco, D. T. Searls, and J. V. Brazie, "Neonatal Mortality Rate: Relationship to Birth Weight and Gestational Age," *Journal of Pediatrics* 81 (1972): 814–22.

Chapter 4: "The Degree of Impairment Is Difficult to Predict"

1. F. Guzzetta et al., "Periventricular Intraparenchymal Echodensities in the Premature Newborn: Critical Determinant of Neurologic Outcome," *Pediatrics* 78 (1986): 995–1006.

2. C. Limperopoulos et al., "Does Cerebellar Injury in Premature Infants Contribute to the High Prevalence of Long-Term Cognitive, Learning, and Behavioral Disability in Survivors?," *Pediatrics* 120 (2007): 584–93.

Chapter 6: Whose Choice?

1. S.N. Wall and J.C. Partridge, "Death in the Intensive Care Nursery: Physician Practice of Withdrawing and Withholding Life Support," *Pediatrics* 99 (1997): 64–70.

2. Leon Eisenberg, "The Human Nature of Human Nature," *Science* 176 (1972): 123–28.

3. R.S. Duff and A.G. Campbell, "Moral and Ethical Dilemmas in the Special-Care Nursery," *New England Journal of Medicine* 289 (1973): 890–94.

4. C.A. Conway, "Baby Doe and Beyond: Examining the Practical and Philosophical Influences Impacting Medical Decision-Making on Behalf of Marginally-Viable Newborns," *Georgia State University Law Review* 25 (2009): 1097–1175.

5. George Annas, "Extremely Preterm Birth and Parental Authority to Refuse Treatment—The Case of Sidney Miller," *New England Journal of Medicine* 351 (2004): 2118–23.

6. In the Matter of Baby K, 16 F.3d F. Supp. 590 (E.D. VA 1993). WL 38674 (4th Cir. 1994).

7. H. MacDonald et al., "Perinatal Care at the Threshold of Viability," *Pediatrics* 110 (2002): 1024–27.

8. J.W. Kaempf et al., "Medical Staff Guidelines for Periviability Pregnancy Counseling and Medical Treatment of Extremely Premature Infants," *Pediatrics* 117 (2006): 22–29.

9. J.W. Kaempf et al., "Counseling Pregnant Women Who May Deliver Extremely Premature Infants: Medical Care Guidelines, Family Choices, and Neonatal Outcomes," *Pediatrics* 123 (2009): 1509–15.

Chapter 7: Is Your Life Good?

1. World Health Organization, "Constitution of the World Health Organization," http://www.who.int/governance/eb/who_constitution_en.pdf.

2. S. Saigal et al., "Parental Perspectives of the Health Status and Health-Related Quality of Life of Teen-Aged Children Who Were Extremely Low Birth Weight and Term Controls," *Pediatrics* 105 (2000): 569–74.

3. S. Saigal et al., "Stability of Maternal Preferences for Pediatric Health States in the Perinatal Period and 1 Year Later," *Archives of Pediatrics and Adolescent Medicine* 157 (2003): 261–69.

4. S. Saigal et al., "Differences in Preferences for Neonatal Outcomes Among Health Care Professionals, Parents, and Adolescents," *Journal of the American Medical Association* 281 (1999): 1991–97.

5. P. Brickman, D. Coates, and R. Janoff-Bulman, "Lottery Winners and Accident Victims: Is Happiness Relative?," *Journal of Personality and Social Psychology* 36 (1978): 917–27.

Chapter 9: Making It Routine

1. N. Charpak et al., "Kangaroo Mother Versus Traditional Care for Newborn Infants </=2000 Grams: A Randomized, Controlled Trial," *Pediatrics* 100 (1997): 682–88.

2. N. Charpak et al., "Kangaroo Mother Care: 25 Years After," *Acta Paediatrica* 94 (2005): 514–22.

Chapter 11: The Plasticity Treadmill

1. J. B. Carmel et al., "Chronic Electrical Stimulation of the Intact Corticospinal System after Unilateral Injury Restores Skilled Locomotor Control and Promotes Spinal Axon Outgrowth," *Journal of Neuroscience* 30 (2010): 10918–26.

INDEX